LOW-CARBON POLICY AND DEVELOPMENT IN TAIWAN

Edited by **Li-Fang Chou and Liang-Feng Lin**

Low-Carbon Policy and Development in Taiwan
Edited by Li-Fang Chou and Liang-Feng Lin

Published by InTech
Janeza Trdine 9, 51000 Rijeka, Croatia

Publishing Process Manager Danijela Duric

Technical Editor Teodora Smiljanic

Cover Designer InTech Design Team

Additional hard copies can be obtained from orders@intechopen.com

Low-Carbon Policy and Development in Taiwan,
Edited by Li-Fang Chou and Liang-Feng Lin
p. cm.
ISBN 978-953-51-0156-7

For EU product safety concerns:
IN TECH d.o.o., Prolaz Marije Krucifikse Kozulić 3, 51000 Rijeka, Croatia,
info@intechopen.com or visit our website at intechopen.com.

Meet the editors

Dr Li-Fang Chou is Director of Green Energy Finance Research Center and Professor of Public Finance at National Chengchi University, Taipei, Taiwan. She received the bachelor degree in economics at National Taiwan University (Taipei), the master degree in public health at National Yang-Ming Medical College (Taipei), and the doctor degree at Institute for Public Finance and Social Policy at Kiel University, Germany. Besides various academic positions in the university, including Director of Center for Public & Business Administration Education and Dean of Office of Research & Development, Dr Chou serves as advisor at several ministerial boards in the government. She has conducted two large-scale national projects in energy economics and management from National Science Council and Ministry of Education.

Dr Liang-Feng Lin is Vice Director of Green Energy Finance Research Center and Associate Professor of Accounting at National Chengchi University, Taipei, Taiwan. He received the bachelor degree in Accounting at National Chengchi University (Taipei), the master degree in Accounting, Chengchi University (Taipei) and Temple University (USA), respectively, and the doctor degree at Economics at Temple University. Besides various academic positions at the university, including Secretary General of Center for Public & Business Administration Education and Director of Learning Research Center, Dr Lin serves as Independent Board Director at South China Insurance Company. He has worked two large-scale national projects with Professor Li-Fang Chou in energy economics and management from National Science Council and Ministry of Education.

Contents

Preface XI

Chapter 1 **Energy Saving and Carbon Reduction Policy in Taiwan 1**
Yi-Cheng Ho and Jenn-Shyong Kuo

Chapter 2 **Renewable Energy Feed-in-Tariff System Design and Experience in Taiwan 25**
Li-Fang Chou and Liang-Feng Lin

Chapter 3 **Assessment of the Decoupling of GHGs and Electricity Costs Through the Development of Low-Carbon Energy Technology in Taiwan 43**
Chien-Ming Lee and Heng-Chi Liu

Chapter 4 **Estimation of Taiwan's CO2 Emissions Related to Fossil Fuel Combustion - A Sectoral Approach 53**
Shinemay Chen, Der-Cherng Lo and Huai Hsuan Yu

Chapter 5 **A Preliminary Look at the Relationship Between Environmental Change and Economic Growth in Taiwan 69**
Kuang-Ta Lo and Ya-Ting Yang

Chapter 6 **Low-Carbon Pilot Tour and Municipal Government Investment: Taiwan's Experience 89**
Liang-Feng Lin

Foreword

With a view to launch a low carbon society in Taiwan by 2050, the Executive Yuan ratified the Sustainable Energy Policy Guide in 2008. Taiwan is now committed to reduce CO2 emission back to its 2005 level by 2020. Then, back to the 2000 level by 2025, and finally achieving 50% of 2000 level by 2050. In addition, the EPA abiding by the Copenhagen Accord to summit Taiwanese NAMAs to the UNFCCC in 2010.

Six authors were involved in putting together this book, and the six papers they wrote were part of the Study of Emissions Trading Scheme Establishment to Respond to Low Carbon and Green Growth in Taiwan program. Founded by the National Science Council, this program was an interdisciplinary analysis of experiences, good practice and progress in the area of low carbon society development in Taiwan.

This book was complied by Professor Li-Fang Chou of Department of Public Finance, National Chengchi University, who is also the Director of Green Energy Finance Research Center. She coauthored the paper "Renewable Energy Feed-in-Tariff Taiwan's Experience." Chien-Ming Lee, of the Institute of Natural Resource Management, National Taipei University, is the author of "Assessment of the Decoupling of GHGs and Electricity Costs through the Development of Low-Carbon Energy Technology in Taiwan." Yi-Cheng Ho of the Department of Public Finance, National Chengchi University, coauthored the paper "Energy Saving and Carbon Reduction Policy in Taiwan." Shinemay Chen of the Department of Public Finance, National Chengchi University, coauthored the paper "Estimation of Taiwan's CO2 Emissions related to Fossil Fuel Combustion–Sectoral Approach." Kuang-Ta Lo of Department of Public Finance, National Chengchi University, contributed an article titled "Environment Change and Economic Growth in Taiwan." Liang-Feng Lin of Department of Accounting, National Chengchi University, who is also the Vice Director of Green Energy Financial Research Center, has been the most important person in the editing process. He also a author of "Low Carbon Pilot Tour and the Investment of Municipal Government."

This book has arrived at the right time, because this is the time to educate the people of Taiwan, about the necessary action for achieving a low carbon society. I am very thankful for this book and would recommend it to the public.

Stephen Shu-hung Shen

Stephen Shen

Minster of the EPA

Preface

Taiwan is a typical small Asian country with few energy resources. Therefore, the question of how to adapt to the trend of a new carbon society has become very important for many developing counties. The present collection of essays not only provides the developmental process of Taiwan's policy, but it also provides an econometric approach to help to estimate CO2 emission levels. The studies also provide some successful examples of how low-carbon regions have helped urban areas revive. Taiwan has become well known for its high-tech industry in the last 20 years. However, Taiwan, as a member of the global village, feels the responsibility to reduce carbon emissions, even though it is not currently an Annex one country. The studies address Taiwan's low-carbon developmental policies of the past 10 years, such as the renewable energy Feed-in-Tariff and the Greenhouse Gas Reduction Act. Besides providing explanations of policy development, the essays also cover an econometric approach to estimate Taiwan's sector department CO2 emissions and to decouple greenhouse gases and electricity costs. The studies further analyze how environmental change affects the economic growth of Taiwan. Finally, the book provides two successful examples of low-carbon pilot regions in Taiwan to explain how a municipal government can use a minimal investment to revive a declining city.

Acknowledgement: The authors and editors of this collaborative effort have received generous funding from the National Science Council (100-3113-P-004-001) in Taiwan. Without the NSC's financial support, the publication of this monograph would not have been possible.

Li-Fang Chou

Liang-Feng Lin

1

Energy Saving and Carbon Reduction Policy in Taiwan

Yi-Cheng Ho[1] and Jenn-Shyong Kuo[2]
[1]*Department of Public Finance, National Chengchi University*
[2]*Department of Accounting, National Taipei University*
Taiwan

1. Introduction

As the international energy situation undergoes sharp changes, greenhouse gas emissions and the safety of energy supplies become the most pressing challenge of energy supply and demand. In this era of the Kyoto Protocol and high oil prices, all countries in the world have put forward reduction strategies for CO2 emissions, including such as developing high-value and low-carbon industrial structures, increasing the energy utilization efficiency of various sectors, and constructing reasonable and effective policy tools for the sustainable development of energy sources, environmental protection and the economy.

As the post-Kyoto Protocol era looms ahead, even though Taiwan is not yet a signatory and is presently free from the pressure of being subjected to a greenhouse gas reduction time limit, as a member of the global village, it still needs to exhibit a sense of responsibility to the international community in protecting the earth. In recent years, Taiwan has referred to the energy balance sheet and the statistical data in the websites of the Environmental Protection Administration and the Ministry of Economic Affairs under the Executive Yuan, and uses the IPCC method to estimate data on greenhouse gas emissions based on reference and sector methods. The Environmental Protection Administration of the Executive Yuan is the present competent authority for the statistics of greenhouse gas emissions, but the statistics of CO2 emissions derived from energy use are estimated by the industry competent authority, the Bureau of Energy of the Ministry of Economic Affairs before being compiled by the Environmental Protection Administration of the Executive Yuan.

Table 1 show that the total greenhouse gas emissions of Taiwan increased gradually from 150 million MT CO2 equivalents in 1990 to 300 million MT CO2 equivalents in 2007, and then decreased to 270 million MT CO2 equivalents in 2009. The CO2 emissions accounted for about 90% which increased from 120 million in 1990 to 270 million MT in 2007. It has been decreasing each year since 2008, and it was 250 million MT in 2009, a decrease of 4.7%. That derived from energy use (fuel combustion) accounted for a large proportion, and the emission was 240 million MT, a decrease of 4.9%.

Unit: $ktCO_2$

Year	CO_2	CH_4	N_2O	HFCS	PFCS	SF_6	CO_2
1990	147,109	122,399	11,974	12,736	NE*	NE	NE
1991	156,609	131,853	11,219	13,537	NE	NE	NE
1992	166,759	141,259	12,116	13,383	NE	NE	NE
1993	181,420	152,725	13,424	13,679	1,592	NE	NE
1994	189,900	160,162	14,000	13,937	1,802	NE	NE
1995	198,445	167,308	15,545	13,902	1,689	NE	NE
1996	208,218	175,754	15,495	14,217	2,752	NE	NE
1997	219,873	188,951	15,447	12,360	3,115	NE	NE
1998	229,788	198,340	15,149	11,908	4,391	NE	NE
1999	237,440	207,130	14,660	12,258	3,392	NE	NE
2000	256,651	224,661	11,028	12,443	5,639	2,386	494
2001	260,193	230,576	9,200	12,437	5,412	2,021	546
2002	267,565	239,593	7,250	12,205	5,415	2,509	593
2003	274,665	248,599	6,196	11,205	4,920	2,776	969
2004	283,565	257,279	5,920	11,734	4,494	2,852	1,285
2005	287,303	263,819	4,979	11,461	1,647	2,505	2,893
2006	294,611	271,774	4,486	11,674	1,028	2,657	2,993
2007	296,801	274,973	4,127	11,429	1,031	2,309	2,933
2008	284,515	263,606	4,727	10,839	1,001	1,498	2,844
2009	272,401	251,149	4,489	10,741	3,619	1,143	1,260

Source: Environmental Protection Administration; Council of Agriculture; Bureau of Energy; Bureau of Industry, ROC

*Note: NE (NOT ESTIMATED)

Table 1. Greenhouse Gas emissions

Table 2 shows the CO_2 emission intensity in Taiwan, the CO_2 emissions per one million NT dollars of real gross product of Taiwan in 2009 was 19.6 MT, a decrease of 0.6 MT as compared with the figure for 2008; the CO_2 emissions per capita were 10.9 MT, a decrease of 0.6 MT.

Year	CO2 Emissions		Emissions Per person	CO_2Emissions Intensity ratios
	Kiloton	Growth rate (%)	(Per-person-kt CO_2)	(kg CO_2/NT$)
1990	122,399	2.10%	6	0.023
1991	131,853	7.72%	6.4	0.023
1992	141,259	7.13%	6.8	0.0229
1993	152,725	8.12%	7.3	0.0232
1994	160,162	4.87%	7.6	0.0226
1995	167,308	4.46%	7.9	0.0222
1996	175,754	5.05%	8.2	0.0221
1997	188,951	7.51%	8.7	0.0225
1998	198,340	4.97%	9.1	0.0229
1999	207,130	4.43%	9.4	0.0225
2000	224,661	8.46%	10.1	0.0231
2001	230,576	2.63%	10.3	0.0241
2002	239,593	3.91%	10.7	0.0238
2003	248,599	3.76%	11	0.0238
2004	257,279	3.49%	11.4	0.0232
2005	263,819	2.54%	11.6	0.0227
2006	271,774	3.02%	11.9	0.0222
2007	274,973	1.18%	12	0.0212
2008	263,606	-4.13%	11.5	0.0202
2009	251,149	-4.73%	10.9	0.0196

Source: Environmental Protection Administration, ROC(2010)
Table 2. CO_2 Emissions from Fuel

The energy supply/demand has grown rapidly in Taiwan over the past two decades. The total CO_2 emissions of fuel combustion in Taiwan in 1990 calculated by the sector method was 122,399 thousand MT; it was 224,661 thousand MT in 2000 and 274,973 thousand MT in 2007, but it decreased to 263,606 thousand MT in 2008 and even further to 251,149 thousand MT in 2009. It decreased by 4.13% in 2008 from 2007, marking the first decrease. It decreased by 4.73% in 2009 from 2008. The figure 1 illustrates the trends of CO2 emissions from 1990 to 2009.

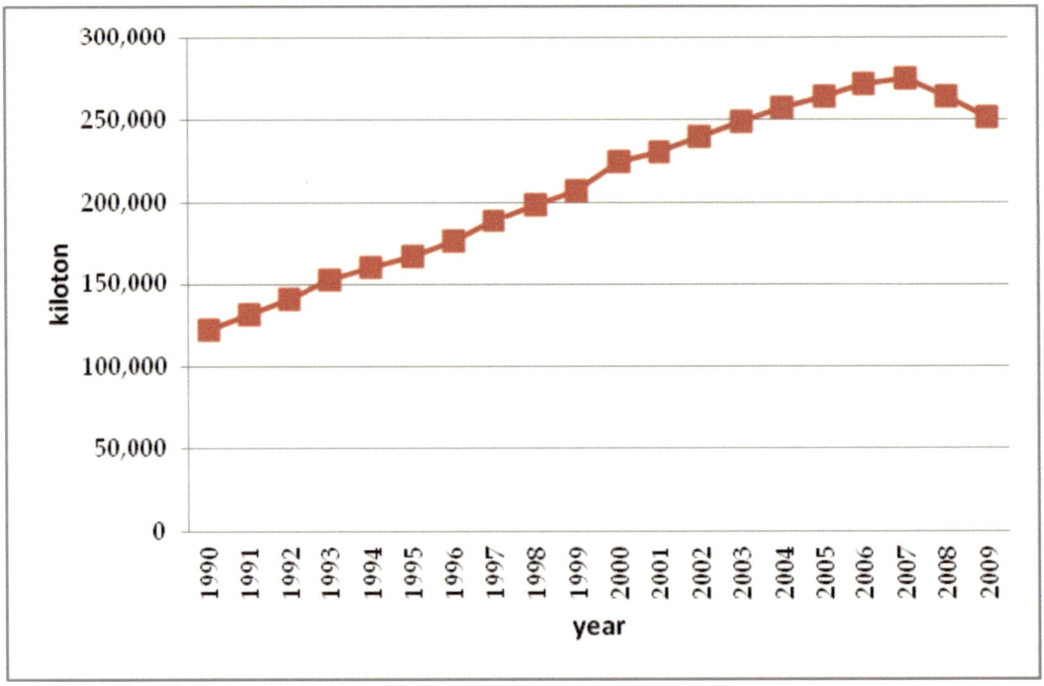

Figure 1. The trends of CO2 emissions from 1990 to 2009

The emission decrease of the most recent years resulted from the financial storm in 2008 which reduced industrial activity, although the energy consumption of various major industries recovered gradually as prosperity revived in 2009. The government has been promoting energy-saving measures since 2008, with energy consumption continuing to evince negative growth. The annual growth rate of 10.29% in 2000 was the highest between 1990 and 2009, followed by the figure of 8.2% in 1991, whereas the positive growth rate of 1.36% in 2007 was the lowest.

According to the data of the Directorate General of Budget, Accounting and Statistics, CO2 emission rate per capita was about 6.0 MT in 1990, 10.1 MT in 2000, and 12 MT in 2007; it decreased to 11.5 MT in 2008, and further decreased to 10.9 MT in 2009. The figure 2 shows the trends of CO2 emissions per capita. The average growth rate of emission per capita between 1991 and 2009 was about 3.4%, but in 2009 it decreased by 5.2% from 2008. In addition, the CO2 emission intensity (i.e., CO2 emission per unit GDP) was 0.023 kg in 1990, 0.0231 kg in 2000, 0.0212 kg in 2007, 0.0202 kg in 2008, and 0.0196 kg in 2009. The emissions in various years and related indexes accounted for about 1% of global emissions, for Taiwan a ranking of 22nd in the world. The figure 3 shows the CO2 emissions intensity form 1990 to 2009.

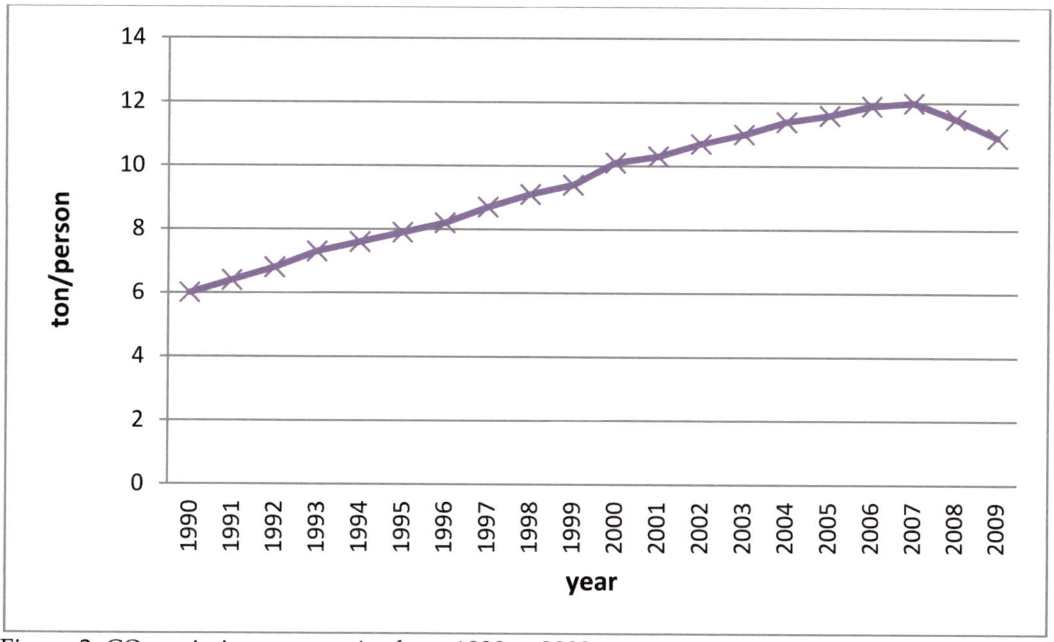

Figure 2. CO_2 emissions per capita from 1990 to 2009

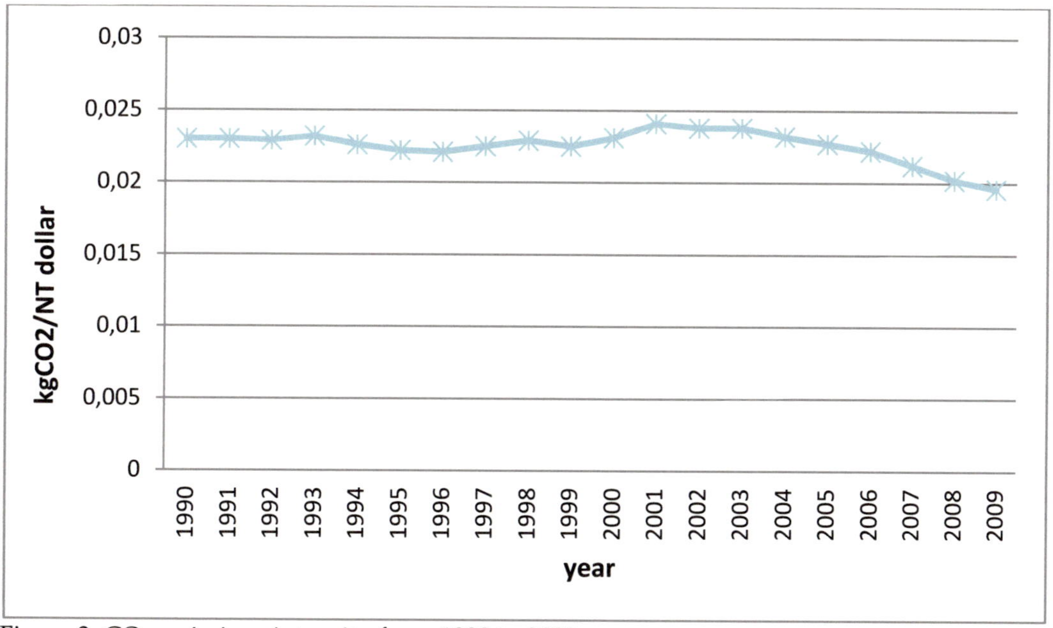

Figure 3. CO_2 emissions intensity from 1990 to 2009

Although energy use decreased as a result of the implementation of various policies in this period, the outcome was not as good as expected for the following reasons: (1) the energy structure has changed, with the proportion of coal with a high carbon content of the overall energy supply continues to increase; (2) the energy price adjustment and tax review policy failed to be implemented effectively, the industrial energy-saving inducement was reduced, and the improvement of energy productivity was obstructed; (3) with respect to energy use and greenhouse gas emission baseline investigation and verification systems, though expected goals were realized, the overall effect still needs to be improved; (4) due to policies promoting the liberalization of the energy industry during the last two decades, many private enterprises in the oil refining and power generation industries have emerged; these enterprises usually use low-cost coal in consideration of their costs, so that the CO_2 emissions have markedly increased.

The fuel combustion CO_2 emission rates of different sectors of Taiwan in 1990 are shown in Table 3 with the energy sector accounting for 45.74%, the industrial sector for 27.26%, transportation for 17.55%, agriculture for 2.63%, housing for 3.59%, and the service industry for 3.23%. In 2009, the energy sector accounted for 65.94%, the industrial sector accounted for 15.90%, the transportation accounted for 13.96%, the agriculture accounted for 0.41%, the housing sector accounted for 2.07%, and the service industry accounted for 1.72%, with the energy sector having the maximum growth rate of energy consumption. Although the carbon emissions of the other sectors all increased, the carbon emission ratios of the sectors other than the energy sector decreased.

Figure 1 shows the data for CO_2 emissions derived from energy consumption in Taiwan for the period 1990-2009. Basically, the emission rate has been increasing linearly since 1990; the annual growth rate is about 11 million MT, even at several key points in time. For example, when the Kyoto Protocol was signed in 1997, and when the Kyoto Protocol went into effect in 2005, the greenhouse gas growth curve of Taiwan continued to develop as before without showing any effect. It is obvious that energy saving and carbon reduction measures undertaken in Taiwan remain inconspicuous.

2. The existing circumstances of energy consumption of various industries in Taiwan

The economic development trend in Taiwan of recent years shows the structural changes of tertiary industry, with the gross product of the industrial sector decreasing year after year, uniting for only 31% of gross product of Taiwan in 2008, whereas the proportion of the service industry has increased continuously, rising to 68% in 2008.

2.1 The structure of energy consumption in Taiwan

The structure of energy consumption in Taiwan is as follows: 98% of Taiwan's energy is imported. Imported petroleum is higher than 99.9%. The energy consumption ratios of different sectors in 2009 are: industry 52.5%, service industry 11.5%, transportation 13.2%, housing 11.6%, energy 7.2%, agriculture 0.9% and non-energy use 3.1%. Figure 5 illustrates the structure of total domestic consumption.

Year	Energy Sector Own Use		Industrial		Transportation		Agricultural		Service		Residential		Total	
	Kt	%	kt	%	kt	%	kt	%	kt	%	kt	%	kt	%
1990	50,705	45.74	30,213	27.26	19,450	17.55	2,917	2.63	3,582	3.23	3,985	3.59	110,851	100
1991	57,187	47.68	31,697	26.43	20,679	17.24	2,673	2.23	3,491	2.91	4,216	3.52	119,943	100
1992	61,268	47.78	33,136	25.84	23,792	18.56	2,646	2.06	2,954	2.30	4,424	3.45	128,220	100
1993	68,944	50.10	33,390	24.26	25,842	18.78	2,648	1.92	2,465	1.79	4,337	3.15	137,626	100
1994	73,930	50.75	34,355	23.58	27,265	18.72	2,694	1.85	2,985	2.05	4,439	3.05	145,669	100
1995	79,925	52.18	34,976	22.83	28,533	18.63	2,749	1.79	2,419	1.58	4,574	2.99	153,176	100
1996	85,546	52.93	35,926	22.23	29,503	18.25	2,776	1.72	3,143	1.94	4,730	2.93	161,624	100
1997	96,476	55.44	37,583	21.60	30,230	17.37	2,451	1.41	2,457	1.41	4,827	2.77	174,024	100
1998	105,773	57.05	38,240	20.63	31,525	17.00	2,021	1.09	2,917	1.57	4,927	2.66	185,403	100
1999	113,262	57.97	39,152	20.04	32,444	16.61	2,020	1.03	3,123	1.60	5,383	2.76	195,384	100
2000	129,737	60.21	42,023	19.50	32,875	15.26	2,338	1.08	3,188	1.48	5,328	2.47	215,488	100
2001	134,875	61.35	40,950	18.63	32,914	14.97	2,430	1.11	3,525	1.60	5,160	2.35	219,855	100

Year	Energy Sector Own Use		Industrial		Transportation		Agricultural		Service		Residential		Total	
2002	138,911	60.97	43,755	19.20	34,197	15.01	2,434	1.07	3,458	1.52	5,081	2.23	227,836	100
2003	149,175	62.89	42,247	17.81	34,164	14.40	2,783	1.17	3,852	1.62	4,992	2.10	237,213	100
2004	155,211	63.27	42,554	17.35	35,501	14.47	2,947	1.20	3,989	1.63	5,101	2.08	245,303	100
2005	161,983	64.36	41,335	16.42	36,478	14.49	2,600	1.03	4,100	1.63	5,203	2.07	251,699	100
2006	169,404	65.34	42,655	16.45	36,406	14.04	1,630	0.63	4,125	1.59	5,046	1.95	259,265	100
2007	173,047	65.85	44,442	16.91	35,071	13.35	1,080	0.41	4,067	1.55	5,080	1.93	262,787	100
2008	167,410	66.42	41,086	16.30	33,103	13.13	1,356	0.54	4,090	1.62	4,997	1.98	252,042	100
2009	158,011	65.94	38,093	15.90	33,447	13.96	994	0.41	4,112	1.72	4,957	2.07	239,615	100

Source: Environmental Protection Administration; Council of Agriculture; Bureau of Energy; Bureau of Industry, ROC
Kt* : Kiloton

Table 3. The fuel combustion CO_2 emissions of different sectors of Taiwan unit: kiloton CO_2 %

2.2 Energy consumption structure by sectors

Energy-intensive industries in the sectoral structure of the manufacturing industry still occupy an important position: energy-intensive industries have a high industry correlation effect, and support the development of other middle and downstream knowledge-intensive industries; they represent a stable raw material supply source for various industries, so they profoundly influence the development of Taiwan's industries. For example, the industrial sector still accounted for 52.5% of energy consumption in 2009. The energy consumption of the industrial sector was 23,145,782 Kl. oil equivalents in 1990, and 59,350,964 Kl. oil equivalents in 2009, an increase of 256% times. The energy consumption of energy-intensive industries was 14,305,778 Kl. oil equivalents in 1990, and 41,040,183 Kl. oil equivalents in 2009, an increase of 286%.

The achievement of voluntary greenhouse gas reduction in energy-intensive industries with assistance of the government has been outstanding in recent years. The six major energy intensive industries jointly signed a greenhouse gas reduction protocol in 2005, expecting to reduce 4.02 million MT CO_2e between 2004 and 2008. The accumulated reduction performance was 3.806 million MT CO_2e between 2004 and 2007, or 1.1% of total emissions.

Over the past 10 years national energy intensity has edged down from 9.43 to 8.82 liters of oil equivalent (LOE) per NT$1000. This shows that the efforts on various fronts to conserve energy have reaped results. But because energy intensive industries (including petrochemicals, steel, textiles and paper) have continued to expand, their 23%growth in energy intensity from 2000 to 2009 has offset improved efficiency in other sectors. The figure 4, 5 and 6 depict the energy consumption by different sectors.

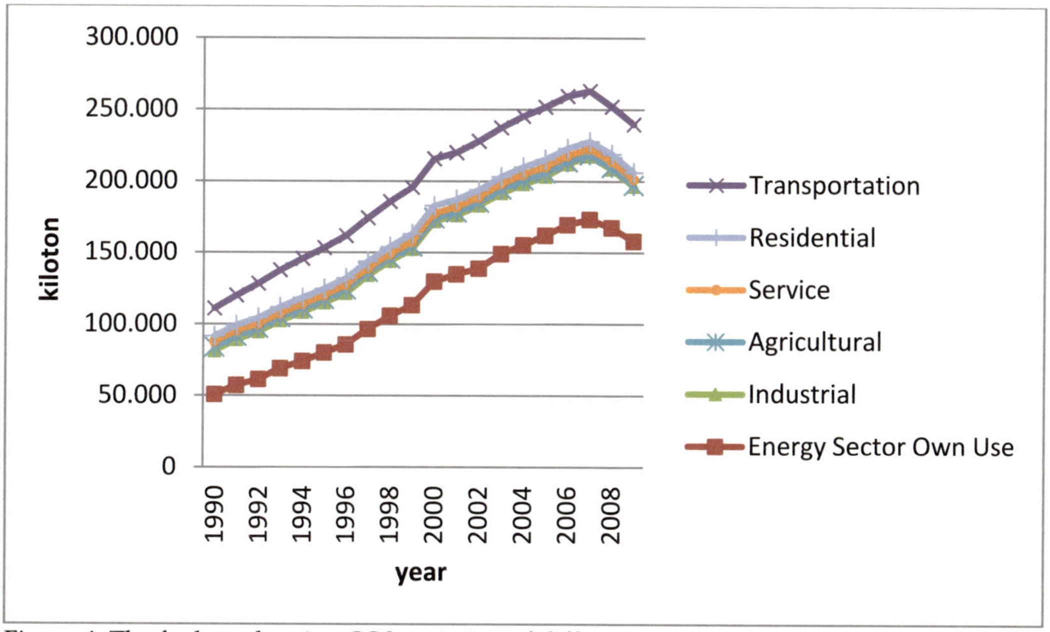

Figure 4. The fuel combustion CO2 emission of different sectors of Taiwan

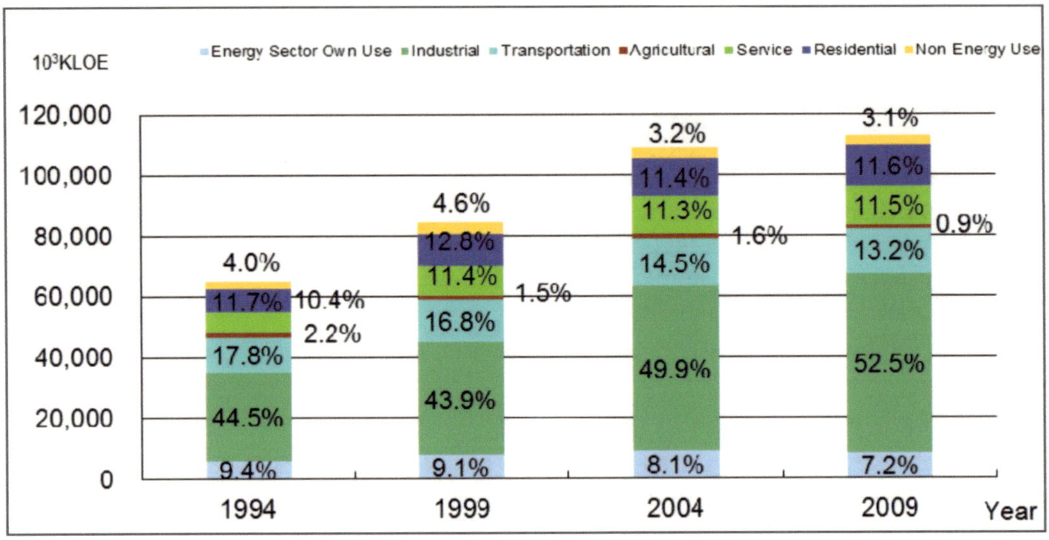

Figure 5. Structure of Total Domestic Consumption (by Sector)

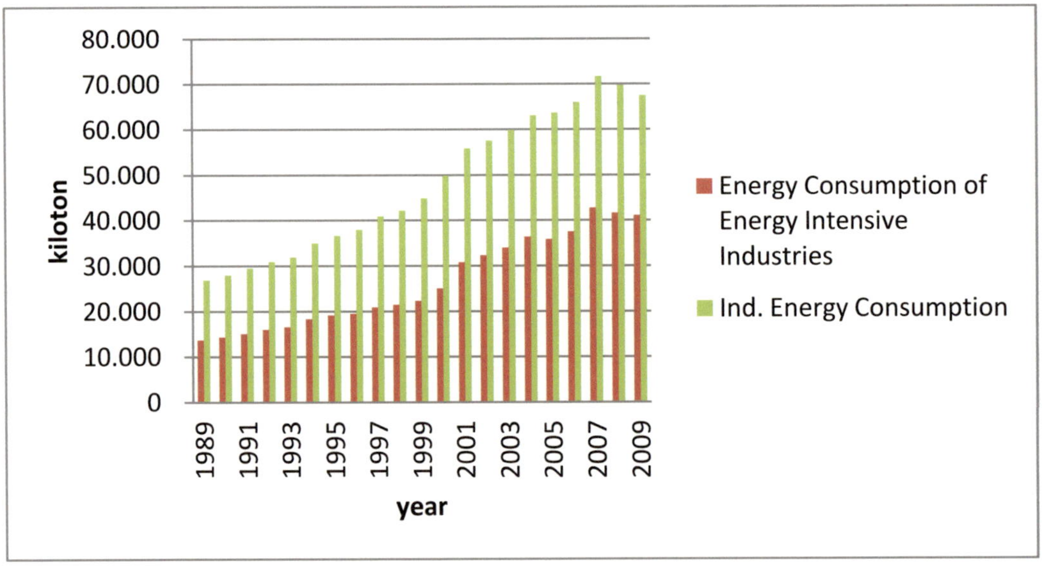

Figure 6. Energy Consumption (by Sector)

3. Current problems in various sectors of Taiwan

International reduction strategies and high oil prices promote energy saving and carbon reduction. A number of advanced countries have developed national reduction strategies based on international commitments since the Kyoto Protocol went into effect. A consensus on reducing at least fifty percent of global greenhouse gas emissions by 2050 was reached at the G8 Summit in July 2008. The international consensus on carbon reduction applies pressure on Taiwan. Taiwan shares the responsibility for reducing emissions and has take position actions in this regard. Furthermore, international oil prices have risen sharply in recent years, and high oil prices have become a long-term trend, making energy efficiency an indicator of inter-industry competitive power. The implementation of measures for energy saving and carbon reduction has come under internal and external pressures.

3.1 Energy efficiency improvement problems in the industrial sector

- The industrial sector has the maximum energy consumption ratio.
- Since the industrial sector has promoted energy saving for a long time, the energy-saving potential of existing equipment is limited.
- The inducement of a voluntary energy conservation agreement is not enough as energy prices are low.
- There are no energy conservation standards of design, construction and the use of business sites and factory buildings.
- There have been no energy efficiency standards of important power equipment (e.g., air compressors, fans, pumps, et al.).

3.2 Energy efficiency improvement problems in the transportation sector

- Transportation demand increases continuously with economic development and population growth.
- Green energy is not yet popular; fuel alternatives are limited.
- Low-carbon transport is difficult to implement in the short term due to high costs.
- The external cost of private transport has not been sufficiently disclosed; the inducement of cost differentials between private transport and public transport remain insufficient.
- The quantity and quality of public transport service still need to be improved.

3.3 Energy efficiency improvement problems in residential and commercial sectors

- Power demand increases continuously with economic development and population growth.
- The energy efficiency of electrical equipment is not clearly indicated.
- The standby power of electrical equipment lacks effective management.
- There is no inducement to invest in green buildings.
- Making energy conservation improvements in old buildings is lacking in incentives.
- No energy conservation standards exist for the design and construction of new buildings.
- High costs make it difficult to equip buildings with renewable energy.
- Inducements for buying energy-saving building materials and appliances still need to be increased.
- The low recovery rate of building materials influences source-reduction performance.

4. Current policy measures in Taiwan

The Executive Yuan of Taiwan adopted the "Sustainable Energy Policy Convention" on June 5, 2008 to construct a "high efficiency", "high value", "low emission" and "low dependence" energy consumption pattern and supply system, so as to realize the three-win vision of energy, environmental protection and economy. The specific measures cover energy saving and carbon reduction of the five major sectors of energy, industry, transportation, environment and life. Regulations and relevant supporting mechanisms have been completed in the hope of attaining the following goals of energy conservation: an increase in the energy efficiency by more than 2% annually to reduce energy use in 2015 by more than 20% from levels in 2005. In terms of carbon reduction the goal is to reduce CO_2 emissions in Taiwan between 2016 and 2020 to levels in 2008, and reduce the levels of emissions in 2025 to those in 2000.

Generally speaking, the energy supply side works on "clean sources" and the energy consumption side works to "reduce expenditures." Energy conversion efficiency must be stressed on the energy conversion side, such as the generating efficiency of power plants and the oil refining efficiency of oil refineries. Higher energy conversion efficiency means "cleaner sources."

Regarding the energy supply side, Taiwan's primary energy supplies are derived mainly from coal, crude oil, natural gas, nuclear energy, and renewable energy. Coal and crude oil are high-carbon energy, whereas natural gas, nuclear energy and renewable energy are classified as low-carbon energy. In terms of clean sources, the ratio of low-carbon energy in the overall primary energy structure must be increased.

Secondly, energy conversion efficiency must be increased, such as increasing the generating efficiency of power plants. High efficiency means using the least primary energy (e.g., coal) to yield the most end-use energy (e.g., electricity) for consumption. Increasing energy conversion efficiency is one of means of developing "clean sources."

Item	Energy saving and carbon reduction mode	Sector	Measures
Energy supply	Clean sources	Energy sector	Adjust energy structure, adopt low-carbon energy (nuclear energy, renewable energy)
		Industrial sector	Adjust industrial structure, encourage low-carbon industries
Energy conversion	Clean sources	Energy sector	Increase generating efficiency of power plants
Energy consumption	Reduce expenditure	Industrial sector	Increase energy utilization efficiency, conserve energy
		Transportation sector	Increase energy utilization efficiency, conserve energy
		Residential and commercial sector	Conserve energy (e.g., ten major measures for energy saving and carbon reduction)
		Government sector	Carbon neutral
		The public	Nationwide energy saving and carbon reduction movement

Table 4. Existing circumstances of the division of work, measures and promotion of energy saving and carbon reduction in Taiwan

Energy consumption is closely related to the daily lives of ordinary people (consumers); it is the starting point from which the general public practices energy saving and carbon reduction, which are the ultimate goals of the "ten major measures for energy saving and carbon reduction" promoted by the government. We usually divide energy consumption into sectors such as transportation, and residential, commercial and industrial sectors for convenience in applying data statistics and policy implementation. This part of energy saving and carbon reduction starts with "reducing expenditures."

The following table summarizes initiatives in carbon reduction and the implementation of primary measures with respect to energy supply, conversion and consumption.

4.1 For "clean sources," reconstruct the energy structure and improve efficiency

- Develop carbon-free renewable energy sources; make effective use of renewable energy development potential, in order to accounts for more than 8% of the generating system by 2025. The total installed capacity of renewable energy is 3.328 million kW, equivalent to 11.1 billion kWh per year, which can reduce about 6.9 million MT CO_2 emissions. The installed capacity of renewable energy is planned to be 8.45 million kW in 2025, accounting for 15.1% of total installed capacity.
    1. Photovoltaic power generation: promote the installation of solar roofs, solar campus, remote off-island emergency disaster prevention, revitalizing the economy; the total installed capacity is 22.4 thousand kW, equivalent to generating 26.91 million kWh per year, so that about 16.7 thousand MT of CO_2 emissions can be reduced.
    2. Wind power generation: the total installed capacity of wind power generation is 518.7 thousand kW (268 units), and the annual power generation is about 1.296 billion kWh, which can serve about 324.2 thousand households and reduce 807.9 thousand MT of CO_2 emissions.
    3. Biodiesel: the estimated annual reduction of CO_2 emission is about 330 thousand MT, equivalent to the CO_2 volume absorbed by about 343 Daan Forest Parks; as for the industrial benefit, there were 10 qualified biodiesel plants up to December 2010, the total annual output is 130 thousand Kl., the accumulated industrial investment of about 1 billion NTD has been driven, when 2% biodiesel is added in, the estimated annual output value is about 3 billion NTD.

- Reduce the carbon footprint of electric power: in order to reduce the CO_2 emissions resulting from the power consumption of other sectors, low-carbon and non-carbon energy generation shall be a primary objective in short-term planning; the efficiency of existing power plants shall be increased in medium-term planning, and the structure of power-generating resources shall be adjusted in long-term planning.
- Improve the overall energy efficiency and energy conservation: energy use was 8.47 liter oil equivalent/thousand NTD in 2010; it was reduced by 3.97% (8.82 liter oil equivalent/thousand NTD) from 2009. Taiwan l energy conservation goal to increase its energy efficiency by more than 2% annually has been attained.

- Accelerate the renewal of power plants; improve the overall efficiency of power plants, and require new power plants to reach an optimal feasible power generation conversion efficiency level consistent with world standards.
- Introduce clean coal technology and developing carbon capture and storage through international research and development; reducing the carbon emissions of generating systems.
- Rationalize energy prices; short-term energy prices reflect internal costs external costs are adjusted progressively in the medium and long term.

4.2 "Reducing expenditure" by promoting substantial energy savings and carbon reduction measures in all sectors

4.2.1 Industrial sector

- Strengthen energy management and increase energy efficiency: according to the "Energy Management Law" passed on July 8, 2009, large-scale productive investment production plans shall be managed in advance, and a mandatory energy label system shall be established.
- Expand energy conservation services: a "comprehensive energy conservation center" is to be established; provide 4,712 energy users with energy conservation guidelines between 2009 and 2012; assist in industrial energy conservation of 525 thousand Kl. oil equivalents (equivalent to reducing energy costs by 10.7 billion NTD).
- Promote voluntary energy conservation in the service industry: a convention at which telecom and communication producers and 3C household appliance groups signed a voluntary energy conservation agreement was held on August 9, 2010. The goal of energy conservation through 2012 is set at 5%. Estimated energy conservation potential is 50 million kWh. Convenience stores, hypermarkets, hospitals, hotels, department stores, supermarkets, shopping centers, telecom and communication producers and 3C household appliance groups of ten major industries (102 group enterprises) signed a voluntary energy conservation agreement effective between 2006 and 2010; the signed groups reduced energy use by 11.9% on average from 2006 to 2009, for a total reduction of 717 million kwh.
- Urge the industrial sector to develop high-added value and low-energy consumption; reduce the carbon emission intensity of unit output value by more than 30% by 2025.
- Check and allocate enterprise carbon credits; assign responsibility for carbon reduction; urge enterprises to promote production and sales systems for energy saving and carbon reduction. Promote voluntary greenhouse gas reduction plans in the energy industry: 33 plants, including Tunghsiao Power Plant of Taiwan Power Co., Ltd. were given assistance in devising voluntary reduction plans up to December 2010; 20 plants were given assistance in gaining approval of their reduction plan designs; 17 plants were given assistance with "ISO 14064-2" verification. Total reduction was more than 6.48 million MT CO_2 equivalents, which shall be used as reference for emission offset or trading of total greenhouse gas control and protecting the preliminary efforts of firms.

- Help small and medium-sized enterprises strengthen their ability to save energy and reduce carbon emissions. Establish inducement measures and management systems and encourage clean production. Energy-saving technology service was provided to 997 enterprises through December 2010. The energy conservation potential of 151.8 thousand Kl. oil equivalent was explored; it was estimated that 80 thousand Kl. oil equivalent could be conserved.
- Encourage popularizing energy saving and carbon reduction and renewable energy and other green energy industries; create a new energy economy.

4.2.2 Transportation sector

- Build seamless urban public transport services; strengthen the accessibility of township public transport systems; take care of remote places.
- Provide real-time traffic information and public transport change information; improve the convenience of public transport and strengthen traffic control functions.
- Build a bicycle path network all over Taiwan; improve the safety, connectivity and continuity of bicycle paths.
- Popularize urban bicycle path networks; provide bicycle parking facilities; establish bicycle rental and riding control systems and facilities; strengthen systematic measures for change in public transport terminal yards.
- Strengthen transport management measures; consider the social cost of private transport in a reasonable manner.
- Internalize the external cost of private transport; promote levying a fuel tax on oil; enlarge the gap of public transport costs.
- Promote ride sharing and safety mechanisms.
- Increase the efficiency level of new cars for private transport by 25% in 2015.
- Promote the reasonable use of biomass fuel in automobiles.
- Promote low-carbon electric vehicles.

Residential and commercial sectors

- Promote mandatory energy efficiency grade labeling: since the announcement of regulations on air-conditioners, refrigerators, cars and motorcycles for energy efficiency grade labeling on July 1, 2010, energy efficiency grade labeling shall be pasted or placed on all such items for sale. Increase the energy efficiency of various power consuming appliances by 10%~70% by 2011, and raise the standard in 2015; popularize high-efficiency products.
- Promote a revolution in energy-saving lighting; promote the "LED traffic signal lamp energy conservation project plan": 17 county and city governments including New Taipei City were given assistance in replacing 135,238 LED traffic signal lamps in 2009 and 2010; 3 municipalities directly under the central government including Taipei City and 11 counties and cities including Keelung City replaced all their signal lamps by 2010. It is estimated that 91% of the 700 thousand traffic signal lamps in Taiwan have been replaced.
- Accelerate the promotion of green buildings; establish systems to encourage their design; assist in providing existing buildings with green building features; provide incentives and rewards.

- Stipulate energy conservation standards of shell energy consumption, air conditioning and lighting systems in the design or construction of new buildings.
- Accelerate the promotion of voluntary agreement of large congregated residential houses; meet the energy conservation potential of the residential sector.
- Provide financial and tax incentives for buying and using green buildings, green building materials and recycled building materials.
- The regulations of buildings shall specify that buildings above a certain scale shall be equipped with renewable energy consuming facilities to increase the use ratio of renewable energy in buildings.

4.2.3 Government sector

- Promote energy conservation in government offices and schools: implement "overall energy saving and carbon reduction measures for government offices and schools"; set negative growth targets for annual power and oil consumption; the overall reduction for 2015 should be 7% of the 2007 figure; help government offices and schools introduce an "energy technology service industry" to improve energy conservation. The energy-saving technology service was implemented in 168 government offices and schools from January to December 2010; the energy potential of 23 thousand Kl. oil equivalents was conserved.
- There shall be a "carbon neutral" concept in policy planning; carbon is to be controlled by putting into practice precautionary, pre-warning and screening principles.

4.2.4 The public

- Promote a nationwide energy saving and carbon reduction movement; encourage the public to "reduce 1 kg carbon footprint per day."
- Promote the policy "discounts for electricity costs for encouraging energy-saving measures": 6,452 households received discounts for electricity costs from July 2008 to January 2011; a total of 10.54 billion kWh was conserved; it was 44% higher than the total power consumption (7.3 billion kWh) of all the households in Taipei City in 2009. The total electric cost deduction was 17.52 billion NTD; CO2 reduction was about 6.70 million MT. The "county-city electricity saving competition" was carried out three times in 2010 since its implementation on July 1, 2010. The first-place winners of the three competitions were Hsinchu City, Chiayi City and Kaohsiung County, respectively. The electricity saving rates was 4.48%, 7.16% and 4.58%, respectively.

5. Taxation tools for energy saving and carbon reduction in Taiwan

There have never been taxes such as a "carbon tax", an "energy tax" or a "green tax" in Taiwan. Taxes related to the environment or energy sources have been levied for the existing policy purposes of energy conservation, environmental protection, maintaining health and rectifying external effects, including an energy tax, a transportation tax, a pollution tax and a tax on resources.

Natural resources may be put in production for economic development, or the manufacturing process may produce wastes or emissions; if the waste of resources and the creation of emissions are not suppressed properly, environmental resources will be exhausted and the environmental quality will deteriorate. For a sustainable utilization of environmental resources, the government can adopt direct administrative control measures for environmental protection and resource management and utilization, and can use economic tools such as an environmental tax, environmental fees, tradable emission permits or quotas, a deposit system and environmental subsidies for environmental protection, so as to carry out the principle by which the environmental media or resource users, or polluters are required to pay fees.

Since the use of environmental taxes tends to be diversified, such taxes benefit both the environment and economy, making it an important policy tool. Acquiring environmental tax data and making comparisons with other countries are feasible steps to take. International organizations are currently discussing the issue of environmental taxes. For example, the OECD makes use of basic statistics; the EU has a statistical handbook; the EU and SEEA have 2003 manuals which define environmental taxes as taxes levied on physical units which have been proved harmful to the environment in a statistical structure. The definition of "tax" is similar to the concept of national income statistics in referring to compulsory and voluntary payments to individual governments.

Environmental tax statistics are divided into four major types in the world, including an energy tax, a transportation tax, a pollution tax and a tax on resources. An energy tax base includes the energy products of transportation and fixed use. A transportation tax is based on the possession and use of motorized vehicles. A pollution tax aims at the discharge of air and water and the management of solid waste and noise. A resources tax aims at water extraction, sandstone, primary raw materials and the exploitation of forest resources; it excludes natural gas and petroleum exploitation (which is regarded as resource rent instead of tax).

Statistical items of the environmental tax in Taiwan include an energy tax, a transportation tax and a pollution tax, but no resources tax. The energy tax includes an energy tariff, oil and gas excise tax and a petroleum fund of energy resources. The transportation tax includes a transport tariff, a vehicle excise tax, a vehicle license tax and a charge for use of automobile fuel. The pollution tax includes air pollution prevention and control fees, soil and ground water contamination regulation fees and recovery and treatment fees.

The present environment-related tax items, competent authorities, sources of law, taxpayers and coverage of taxation in Taiwan are shown in Table 5.

Table 6 illustrates the environmental tax rates in Taiwan. Table 7 shows the environmental tax revenues and composition of Taiwan; it also shows that tax revenues totaled 227.89 billion NTD in 2009, an increase of 0.2% from 2008. The transportation tax of 132.12 billion NTD accounted for 58.0%; the energy tax of 84.86 billion NTD accounted for 37.2% for a combined total of 95.2%. The pollution tax was 10.9 NTD, accounting for 4.8%.
The energy tax was 84.86 billion NTD in 2009, an increase of 9.8% from 2008; the oil gas excise tax of 84.29 billion NTD accounted for 99% of the energy tax. The petroleum fund was

330 million NTD, accounting for only 0.4%. In addition, imported energy decreased greatly as a result of the economic recession. Revenue from the petroleum tariff decreased to 230 million NTD, a sharp decrease of more than 75%; its proportion decreased to 0.3%.

Revenue from the transportation tax was 132.12 billion NTD, a decrease of 4.4% from 2008. Revenue from the vehicle license tax was 53.05 billion NTD, accounting for 40.2% of the transportation tax. Automobile fuel fees totaled 43.24 billion NTD, accounting for 32.7%; in addition, the vehicle excise tax decreased 17.6% from 2008 as a result of poor motorcycle sales in Taiwan; its proportion decreased to 21.0%.

The pollution tax generated revenues of 10.9 billion NTD in 2009, a decrease of 9.7% from 2008. The "Recycling, Clearance, and Disposal Fees" resulted in 6.01 billion NTD, accounting for 55.1% of pollution tax revenues. The "Air Pollution Control Fee "in Pollution Control accounted for 39.3%, and the "Soil and Groundwater Pollution Remediation Fee "oil and Groundwater accounted for only 5.6%.

Tax item	Tariff	Excise tax	Vehicle's Fuel Charge	Air pollution fee	Soil pollution fee	Petroleum fund
Competent authority	Ministry of Finance	Ministry of Finance	Ministry of Transportation and Communications	Environmental Protection Administration	Environmental Protection Administration	Ministry of Economic Affairs
Source of law	Customs Law	Excise Tax Regulations	Highway Law	Air Pollution Control Law	Soil and Groundwater Pollution Remediation Law	Petroleum Management Act
Taxpayer and coverage of taxation	Imported dutiable goods shall be levied on according to tariff regulations.	Taiwan made or imported dutiable goods shall be levied on according to excise tax regulations.	All vehicles on highways or in urban area, excluding the tax-exempt vehicles specified in Article 4 of Highway Law.	Mobile pollution sources: the distributors or users are levied on according to the variety and quantity of discharged air pollutants, and the distributors or importers are levied on according to the composition and quantity of oil fuels.	Makers and importers of specified chemical substances are levied on according to the output quantity and input quantity.	Petroleum products importers, explorers or manufacturers are levied on.

Source: Ministry of Finance, Directorate General of Budget, Accounting and Statistics, Executive Yuan, R.O.C (2011).

Table 5. Taiwan's Environment-related Taxes

Taiwan's taxation policies for various environment-related taxes in the future: The "Regulation for Energy Tax (draft)" is being scheduled for legislative review.

The fundamental aspects of the Energy Tax are as follow:

- Implement an energy tax to maintain a financial balance.
- According to the conclusions of the Tax Reform Committee, Executive Yuan, the implementation of a green tax system will integrate the present oil gas excise tax, automobile fuel fee and petroleum fund. The increased tax revenues after implementation will be used to subsidize low-income households and public transport first; untimely items of excise tax, stamp duty and amusement tax systems will be reformed to reduce their impact on the public.
- As the energy tax system will exert a significant influence on industry and the economy, the Ministry of Finance will study the influence of the energy tax on the economy, industry and the environment as well as the opinions of all circles of society before drafting and planning the tax system, which will be implemented at a proper time.

Item	Import Tariff			Trade Promotion Service Fee	Excise Tax	Petroleum Fund	Soil and Groundwater Pollution Remediation Fee	Air Pollution Control Fee	Value Added Tax
	Column I	Column II	Column III						
Crude Oil	0%	0%	2.50%	0.04%	(Free)	109 NT $ /KL	0NT $ /T	0	5%
Fuel Oil	5%(2.5%)	0%	5%	0.04%	0.11NT $ /L	137 NT $ /KL	12NT $ /T	0	5%
Kerosene	0%	0%	15%	0.04%	4.25NT $ /L	133NT $ /KL	0NT $ /T	0	5%
Kerosene Type Jet Fuel	0%	0%	15%	0.04%	0.61NT $ /L	133NT $ /KL	0NT $ /T	0	5%
LPG	0%	0%	2.50%	0.04%	0.69NT $ /KG	151NT $ /T	(Butane)12NT $ /T	0	5%
Gasoline	0%	0%	15%	0.04%	6.83 NT $ /L	169 NT $ /KL	22NT $ /T	1st Grade 0.03 NT$/L 2nd Grade 0.075 NT$/L 3rd Grade 0.19 NT$/L	5%
Diesel Oil	0%	0%	15%	0.04%	3.99NT $ /L	144NT $ /KL	22NT $ /T	1st Grade 0.03 NT$/L 2nd Grade 0.075 NT$/L 3rd Grade .2 NT$/L	5%
Natural Gas	0%	0%	7.50%	0.04%	0	0	0	0	5%
Steam Coal	0%	0%	0%	0.04%	0	0	0		5%
Coking Coal	0%	0%	0%	0.04%	0	0	0		5%
Electricity	-	-	-	-	0	-	0		5%

Source: Ministry of Finance, Directorate General of Budget, Accounting and Statistics, Executive Yuan, R.O.C (2011).
Note: The import tariff rate is divided into three columns. The first column applies to goods imported from WTO members or from countries or areas that have reciprocal agreements with the Republic of China. The second column applies to specified goods imported from specific underdeveloped or developing countries or areas, or from countries or areas which have signed a Free Trade Agreement with the Republic of China. When the rates in the first and second column are not applicable, the rates in the third column shall apply.

Table 6. Energy-Related Tax Rates

Unit: million NTD;%

Year	2007		2008		2009	
	amount	%	**amount**	%	**amount**	%
Total	**253,801**	**100**	**227,517**	**100**	**227,886**	**100**
Energy Tax	**89,324**	**35.19%**	**77,290**	**33.97%**	**84,859**	**37.24%**
Import Tariff	910	0.36%	964	0.42%	233	0.10%
Oil Gas Excise Tax	87,560	34.50%	75,735	33.29%	84,293	36.99%
Petroleum Fund	854	0.34%	591	0.26%	333	0.15%
Transportation Tax	**152,219**	**59.98%**	**138,157**	**60.72%**	**132,124**	**57.98%**
Import Tariff	10,816	4.26%	7,419	3.26%	8,091	3.55%
Vehicle Excise Tax	43,953	17.32%	33,677	14.80%	27,741	12.17%
Vehicle License Tax	53,271	20.99%	53,255	23.41%	53,050	23.28%
Vehicle Fuel Charge	44,179	17.41%	43,806	19.25%	43,242	18.98%
pollution tax	**12,258**	**4.83%**	**12,070**	**5.31%**	**10,903**	**4.78%**
Air Pollution Control Fee	4,810	1.90%	4,946	2.17%	4,282	1.88%
Soil and GW Pollution RD Fee	711	0.28%	575	0.25%	612	0.27%
Recycling, Clearance, Disposal	6,737	2.65%	6,549	2.88%	6,009	2.64%

Source: Ministry of Finance, Directorate General of Budget, Accounting and Statistics, Executive Yuan, R.O.C (2011).

Table 7. Energy-related Tax Revenues and Composition

6. Conclusion

Taiwan is encountering the global environmental crises related to global warming and faces continuing challenges from the environmental deterioration stemming from economic development. Taiwan is deficient in conventional energy resources and highly dependent on energy imports, with nearly 90% of its greenhouse gas (GHG) emissions coming from carbon dioxide emitted from energy use. The annual growth of GHG emissions has been slowing in recent years, with negative growth reported for the first time in 2008. Taiwan is currently not a signatory to the United Nations Framework Convention on Climate Change; however, as a member of the global village Taiwan has committed itself to sharing the obligations of common but differential responsibility in accordance with the basic principles of the UNFCCC. In order to maintain national competitiveness and limit the consumption of high-priced energy, the government will continue to implement energy conservation and carbon reduction measures for national sustainable development.

But if we truly want to reshape the nation's industrial structure through imposed controls, effective means should include both the imposition of energy taxes and the passing of a Greenhouse Gas Emissions Reduction Act. Energy conservation and carbon reduction in Taiwan will never be just a slogan, but a new lifestyle in action.

7. References

[1] Action for a low-carbon society : Northern Ireland," Energy Saving Trust, London Press, 2007.
[2] Advanced Energy Initiative," National Economic Council, February 2006, http://www.whitehouse.gov/stateoftheunion/2006/energy/energy_booklet.pdf
[3] Building a Low-Carbon Society," Ministry of the Environment, 2007. http://www.env.go.jp/earth/info/pc071211/en.pdf
[4] 12.U.S. Centers for Disease Control and Prevention, Active Transportation to School Then and Now, Barriers and Solutions.
[5] 2005 National Energy Conference, Bureau of Energy, Ministry of Economic Affairs. http://www.moeaboe.gov.tw/hot/EnergyMeeting/defalult.htm
[6] An interview with Dr Reid Ewing, Relationship between urban sprawl and physical activity, obesity, and morbidity, American Journal of Health Promotion, 18: 47-57, 2003.
[7] Bureau of Energy, Ministry of Economic Affairs, 2009 "Energy Statistics" http://www.moeaboe.gov.tw/English/Statistics/EnStatistics.aspx
[8] Energy Policy Act of 2005, Public Law 109-58—Aug. 8, 2005, Congressional Record, Vol. 151, 2005, http://www.epa.gov/oust/fedlaws/publ_109-058.pdf
[9] Energy Statistics Handbook 2008, Bureau of Energy, Ministry of Economic Affairs, Taiwan; 2008.
[10] Ewing, R and R Cervero, Travel and the Built Environment: A Synthesis, Transportation Research Record, 1780, 87-114, 2001.
[11] Executive Yuan, ROC 2008, "Sustainable Energy Policy Convention" http://www.ey.gov.tw/ct.asp?xItem=60604&CtNode=3434&mp=95
[12] Future Gen, http://www.fossil.energy.gov/programs/powersystems/futuregen/
[13] Hydrogen Fuel Initiative, http://www.hydrogen.gov/thepresidentshydrogen_fi.html
[14] IPCC Fourth Assessment Report, Climate Change, 2007, 2008

[15] Jim S. & N. Shuzo, "Policies and practices for a low-carbon society," *Climate Policy* 8, 5-16, 2008.
[16] Ken Worthy JR. "Transport Energy Use and Greenhouse Emissions in Urban Passenger Transport Systems, A Study of 84 Global Cities." http://cst.uwinnipeg.ca/documents/Transport_Greenhouse.pdf
[17] Meeting the Energy Challenge, *"A White Paper on Energy"*, May 2007, Department of Trade and Industry, UK
[18] Ministry of Economic Affairs, *"Framework of Taiwan's Sustainable Energy Policy"*, 2008.
[19] Renewable Fuel Program, http://www.epa.gov/otaq/renewablefuels/
[20] Social Indicators 2008, Directorate General of Budget, Accounting and Statistics, Executive Yuan, Taiwan. http://eng.dgbas.gov.tw/.
[21] Towards a European Strategic Energy Technology Plan(SET Plane), *Commission of the European Communities,* Brussels, 10.1.2007, COM(2006) 847; also in *IEA Energy Policies Review*: The European Union 2008, ISBN: 978-92-64-04337-4
[22] W.M. Huang and G.W.M. Lee, Feasibility analysis of GHG reduction target: lessons from Taiwan's energy policy, *Renewable and Sustainable Energy Reviews* **13** (2009), pp. 2621-2628.

2

Renewable Energy Feed-in-Tariff System Design and Experience in Taiwan

Li-Fang Chou[1] and Liang-Feng Lin[2]
[1]*Department of Public Finance, National Chengchi University*
[2]*Department of Accounting, National Chengchi University*
Taiwan

1. Introduction

Facing climate change, energy dependency and energy security and other significant environmental challenges, many countries try to seek environmental sustainability, promote a green new deal, and develop renewable energy. IPCC (2011) found that building a low-carbon city, developing low-carbon industry and promoting low-carbon life are the major means for most countries to achieve a low-carbon society.

The major renewable energy sources include solar energy, wind power, biomass, geothermal, hydro power et al. REN21 (2011) Renewables 2011 Global Status Report indicated that in 2009 global renewable energy sources supplied 16% of global final energy consumption. In 2011 additional investments of renewable energy in the world were US$211 billion and the top 5 new capacity investment countries were China, Germany, the United States, Italy, and Brazil, respectively. In terms of new investment in types of energy, China was among the top-ranking countries in wind power and solar heat; Germany was at the top in solar photovoltaic and biodiesel production sources, and the United States was tops in ethanol production.

In 2010 the worldwide total renewable energy capacity was 1,320 gigawatts (GW), and the largest 3 types of renewable energy capacity (REC) were hydro power 1,010 GW, wind power, 198GW and the energy PV 40GW. The top 5 countries of REC were China, the United States, Canada, Brazil, and Germany/India. China was ranked at the top in capacity of wind power and solar heat; the United States was ranked first in biomass and geothermal power, and Germany was number one in solar PV (REN21 2011).

In recent years the two most important renewable energy tools in the European Union (EU) have been the Feed-in Tariff (FIT) and the Quota/TGC (a quota regulation in combination with a tradable green certificate). Twenty out of twenty-seven EU member nations are using FIT as their main renewable energy tool (Klein, et al., 2008). Table 1 reveals that no matter the extent of economic growth or national income distribution, all countries in the world promote an FIT policy to deal with the impact of environmental change (REN21, 2011). Taiwan is located in a sub-tropical area with abundant sunshine, surrounded by seas with strong wind power and ample currents; therefore, the island is suitable for developing renewable energy (Bureau of Energy, Ministry of Economic Affairs, 2009). Taiwan has implemented an FIT policy since 2009. The purpose of the policy is not only to develop renewable energy aggressively, but also to save energy, reduce carbon emissions as well as ease the threat of excessive dependency on energy imports.

A renewable FIT policy focuses on two objectives: an Access Objective and a Price Objective (PACT, 2011). The Access Objective implies that the local power company that uses renewable energy power generation equipment to produce electricity and operate a power grid shall have the obligation to provide parallel connections and wholesale rates. The system design focuses on ensuring a connection to the grid, extending and reinforcing the grid and sharing reasonable costs. The Price Objective emphasizes setting a tariff at a reasonable level, guaranteeing a price for a designated period of time, and offering a reasonable return on investment. The system focuses on a tariff (price), a wholesale period and a wholesale rate, and an adjustable mechanism are very important, too (PACT 2011, Mendonça, Jacobs and Sovacool, 2010, and Chou, Lin, and Chen 2010)

<table>
<tr><td rowspan="2">High-income countries</td><td>National level policy</td><td>Austria, Croatia, Cyprus, Czech Republic, Demark, Estonia, Finland, France, Germany, Greece, Hungary, Ireland, Israel, Italy, Japan, Latvia, Luxembourg, Portugal, Slovakia, Slovenia, Spain, Switzerland, United Kingdom</td></tr>
<tr><td>No national level policy</td><td>Australia, Canada, United States</td></tr>
<tr><td>Upper-middle income countries</td><td colspan="2">Algeria, Argentina, Bosnia & Herzegovina, Bulgaria, Costa Rica, Dominican Rep., Kazakhstan, Lithuania, Macedonia, Malaysia, Panama, Peru, Serbia, South Africa, Turkey</td></tr>
<tr><td>Lower-middle income countries</td><td colspan="2">Armenia, China, Ecuador, Honduras, India, Indonesia, Moldova, Mongolia, Nicaragua, Philippines, Sri Lanka, Thailand, Ukraine</td></tr>
<tr><td>Low-income countries</td><td colspan="2">Kenya, Tanzania, Uganda</td></tr>
</table>

Source: REN21 (2011), Renewables 2011: global status report, http://www.ren21.net/
Taiwan started to provide Feed-in-Tariff countries in 2009.

Table 1. Feed-In Tariff Countries

This chapter attempts to analyze the renewable energy FIT system design and practice in Taiwan. First, the chapter indicates the present status of energy consumption and supply in Taiwan; then it introduces Taiwan's Renewable Energy Development Law (REDL); a discussion of the financial mechanism of the FIT in Taiwan follows; finally, we examine the effectiveness of Taiwan's FIT. In the meantime, we also want to introduce Taiwan's FIT model to members of the international academic community who are interested in related topics.

2. The Present Status of Energy Supply and Demand in Taiwan

2.1 Energy Consumption

The population of Taiwan is more than 23 million. The growth of energy consumption was very rapid; from 1990 to 2010 the annual growth rate was 4.39%. Energy consumption in 1990 was 50.99 million kiloliters of oil equivalent (KLOE) and in 2010 was 120,308 KLOE. Per capita energy consumption in 1990 was 2,520 liters of oil equivalent (LOE) and the number increased to 5,223 LOE in 2010. Table 2 lists the economic indicators for energy.

Item / Year	Mid-Year Population (1,000 Persons)	Total Domestic Consumption Quantity (10^3KLOE)	Per Capita Energy Consumption (LOE)
1990	20,233.00	50,986.70	2,519.98
1995	21,215.00	68,472.50	3,227.55
2000	22,125.00	91,737.40	4,146.32
2005	22,652.40	111,168.30	4,907.57
2010	23,035.40	120,308.00	5,222.74

Note: 1. Domestic Consumption = Energy Sector Own Consumption + Final Consumption 2. Final consumption = Industrial Sector + Transportation Sector + Agricultural Sector + Services Sector + Residential Sector + Non-Energy Consumption
Source: Bureau of Energy, Ministry of Economic Affairs (2010).

Table 2. Economic Indicators for Energy

Table 3 illustrates energy consumption by sector in 1990 and 2010. Consumption in the energy sector was 6.97%, industrial sector 53.81%, transportation sector 12.91%, services sector 10.95%, residential sector 10.71%, agricultural sector 0.8% and non-energy use 3.8%. Table 3 reveals energy consumption by source in 1990 and 2010. Consumption of coal and coal products was 8.33%, petroleum products 40.23%, natural gas 2.46%, electricity 48.60%, solar thermal sources 0.09%, and heat 0.29% (Bureau of Energy, Ministry of Economic Affairs, 2010). For the same period electricity consumption increased 6.15% from 1990 to 2010, but oil consumption decreased 5.57%, which demonstrates that electricity consumption showed an upward trend while oil consumption showed a downward trend in the last decade.

By Sector	1990	2010
Non-Energy Use	5.13%	3.83%
Residential	11.66%	10.71%
Services	9.75%	10.95%
Agricultural	2.86%	0.82%
Transportation	15.71%	12.92%
Industrial	45.40%	53.81%
Energy Sector Own Use	9.50%	6.97%

Source: Bureau of Energy, Ministry of Economic Affairs (2010).

Table 3. Structure of Total Domestic Consumption (By Sector)

Unit:%

Year	Total	Coal & Coal Products	Petroleum Products	Natural Gas	Electricity	Solar Thermal Power	Heat
1990	100.00	9.10	45.80	2.61	42.45	0.04	-
1995	100.00	7.36	44.57	3.54	44.44	0.08	-
2000	100.00	7.06	39.90	2.58	50.37	0.08	0.00
2005	100.00	6.68	41.17	2.07	49.88	0.09	0.11
2010	100.00	8.33	40.23	2.46	48.60	0.09	0.29

Source: Bureau of Energy, Ministry of Economic Affairs (2010).

Table 4. Total Domestic Consumption (by Energy Form)

2.2 Energy supply

Table 5 illustrates that the energy supply increased almost 2.5 times from 1990 to 2010 and the total amount increased from 58.52 million KLOE to 145.56 million KLOE. The annual growth rate was 4.66% from 1990 to 2010. In 2010 the indigenous energy of Taiwan only accounted for 0.61% of the total energy, and imported energy accounted for 99.39%. The indigenous energy included crude oil (0.01%), natural gas (0.18%), conventional hydro power (0.28%), solar photovoltaic and wind power (0.07%), and solar thermal power (0.08%). Imported energy included crude oil and petroleum products (49.03%), coal and coal products (32.09%), liquid natural gas (9.98%) and nuclear power (8.28%).

Unit: 10^3KLOE

Item	2010		1990		Average Growth Rate %
	Quantity	%	Quantity	%	
Total Supply	145,560.90	100.00	58,520.7	100.00	4.66
By Indigenous & Imported					
Indigenous Energy	893.0	0.61	2,313.8	3.95	-4.65
Coal	-	-	325.2	0.56	-100.00
Crude Oil	14.2	0.01	182.4	0.31	-11.97
Natural Gas	263.3	0.18	1,173.9	2.01	-7.20
Conventional Hydro Power	401.0	0.28	610	1.04	-2.08
Solar Photovoltaic and Wind Power	100.2	0.07	2.7	0.00	19.86
Solar Thermal Power	114.3	0.08	19.6	0.03	9.21
Imported Energy	144,667.9	99.39	56,206.9	96.05	4.84
Coal & Coal Products	46,710.9	32.09	13,696.1	23.40	6.33
Crude Oil & Petroleum Products	71,375.5	49.03	32,137.2	54.92	4.07
Liquid Natural Gas	14,525.8	9.98	855.6	1.46	15.21
Nuclear Power	12,055.7	8.28	9,518.0	16.26	1.19

Source: Bureau of Energy, Ministry of Economic Affairs (2010).

Table 5. Taiwan's Energy Supply in 1990 and 2010

2.3 Energy Efficiency and Security

Table 6 shows the energy efficiency indicators of Taiwan from 1990 to 2010. Taiwan's Energy productivity (Real GDP / Domestic Energy Consumption) in 1990 was NT $104 /LOE, and in 2010 the number increased to NT $118 /LOE. The energy intensity of Taiwan (Domestic Energy Consumption / Real GDP) was 9.95 LOE/ NT $1,000 in 1990, and in 2010 the number went up to 8.46 LOE/ NT $1,000. Per capita electricity consumption in 1990 was 4,193 KWh and in 2010 the number increased to 10,312 KWh. The average electricity price was NT $2.1636 / KWh in 1990 and the price went up to NT $2.6098 / KWh (Bureau of Energy, Ministry of Economic Affairs 2010). The numbers indicate that energy efficiency in Taiwan increased from 1990 to 2010 gradually.

Item / Year	Energy Productivity (NT$/LOE)	Energy Intensity (LOE/NT$1,000)	Per Capita Electricity Consumption (KWh)	Average Electricity Prices (N.T.$/KWh)
1990	104.27	9.59	4,193.49	2.1636
1995	110.06	9.09	5,940.95	2.1859
2000	106.08	9.43	7,978.51	2.1133
2005	104.46	9.57	9,651.20	2.0533
2010	118.15	8.46	10,312.80	2.6098

Source: Bureau of Energy, Ministry of Economic Affairs (2010).

Table 6. Energy Efficiency Indicators

Table 7 illustrates Taiwan's energy security indicators. Taiwan's dependence on imported energy in 1990 was 96%; the value of energy imports to the value of total imports was 11.45%; the value of energy imports to GDP was 3.8%; dependency on imports oil was 99.43%, and the amount of per capita energy imports was NT $8,328. In 2010 those numbers increased to 99.3%, 20.06%, 11.74%, 99.97%, and NT $69,317, respectively.

2.4 Electricity Rate Adjustment

In order to reflect power generating and purchasing cost and the fluctuation of international electricity prices in a timely manner, the Ministry of Economic Affairs (MOEA) approved "Taipower Electricity Price and Fuel Mechanism" (TEPFM) of Taipower. The mechanism stipulates that the all basic electricity price rates shall be based on the rate as of October 1, 2008. The mechanism also allows Taipower to adjust the price when the average unit cost of fossil fuels (gas, coal, and oil fuel) changes and the electricity price could follow accordingly. Taipower also needs to publicly announce "the actual weight average fuel cost per KW" and "the relative weight average fuel cost per kWh" on a quarterly basis and file a report with the MOEA. According to TEPFM, the time to initiate an electricity price adjustment is as follows: when an absolute dollar amount per kWh electricity cost has grown (decreased) more than 1 percent of the average electricity selling price per kWh of the first half year average, then Taipower could initiate the price adjustment mechanism which allows

Taipower to add (deduct) "a fuel adjustment unit cost per kWh " to the basic electricity rate. If it is less than 1 percent of "the average electricity selling price per kWh" of the first half year average then the addition (deduction) shall be canceled.

Item / Year	Dependence on Imported Energy	Value of Energy Imports/ Value of Total Imports (%)	Value of Energy Imports/ GDP (%)	Dependence on Imports Oil (%)	Value of Oil Imports/ Values of Total Imports (%)	Value of Oil Imports / GDP (%)	Per Capita Energy Imports (NT$)
1990	96.01	11.45	3.80	99.43	9.14	3.03	8,328
1995	97.97	6.86	2.58	99.85	4.98	1.87	8,867
2000	98.74	9.03	3.88	99.93	7.08	3.04	17,875
2005	99.15	16.02	7.94	99.94	12.27	6.08	41,151
2010	99.30	20.06	11.74	99.97	14.90	8.72	69,317

Source: Bureau of Energy, Ministry of Economic Affairs (2010).

Table 7. Energy Security Indicators

After analyzing the range of Taiwan electricity price adjustments, in 1990 the average price per kWh was NT$2.1636; in 2010, the average price per kWh was NT$2.1636 and the growth rate was 20.62%. During the same period, the national income per capita, representing economic ability, was US$7,628 in 1990, and US$16,432 in 2010, and the growth rate was 115.41%. The consumer price index, representing living expenses, was 74.49% in 1990 and 105.48% in 2010, and the growth rate was 41.60%. The above comparison indicates that the range of electricity price increases is much narrower than those of national income per capital and the consumer price index. Therefore, whether the electricity price is fully reflected in energy costs and whether the external costs are reasonably internalized will affect the relative price, using the motivation of renewable energy as well as the willingness to invest in renewable energy equipment.

Year	Average electricity price (N.T$/kWh)	Per capita of national income	Consumer price index
1990	2.1636	7,628	74.49
1991	2.1629	8,473	77.18
1992	2.1847	9,843	80.63
1993	2.1943	10,244	83.00
1994	2.1851	11,068	86.41
1995	2.1859	11,882	89.58
1996	2.1905	12,330	92.33
1997	2.1575	12,652	93.17
1998	2.1605	11,419	94.73

1999	2.1071	12,279	94.90
2000	2.1133	13,299	96.09
2001	2.1221	11,821	96.08
2002	2.0945	12,077	95.89
2003	2.0682	12,549	95.62
2004	2.0520	13,602	97.17
2005	2.0533	14,412	99.41
2006	2.1046	14,724	100.00
2007	2.1484	15,192	101.80
2008	2.3010	15,194	105.39
2009	2.6070	14,271	104.47
2010	2.6098	16,432	105.48

Note: 2006 is the base year for the consumer price index.

Sources: 1. Bureau of Energy, MOEA (2011a), Energy Statistical Annual Reports.2. Directorate General of Budget, Accounting and Statistics (DGBAS) of Executive Yuan http://www.dgbas.gov.tw/ct.asp?xItem=393&CtNode=2850&mp=1

Table 8. Electricity Prices and Economic Index

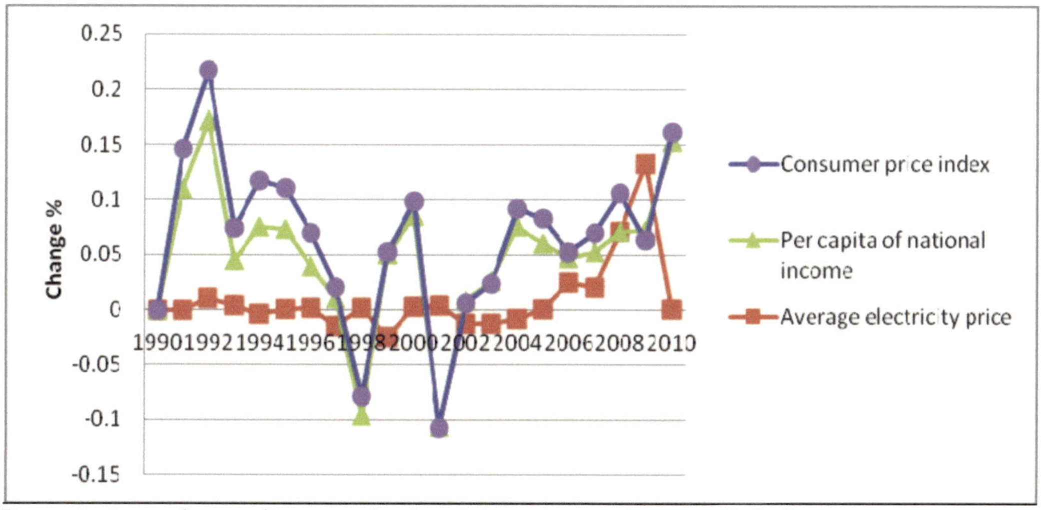

Figure 1. Annual rate of increase in electricity prices and economic index
1. Bureau of Energy, MOEA (2011a), Energy Statistical Annual Reports.
2. Directorate General of Budget, Accounting and Statistics (DGBAS) of Executive Yuan http://www.dgbas.gov.tw/ct.asp?xItem=393&CtNode=2850&mp=1

3. Introduction of Taiwan's Renewable Energy Development Law

In 2008 the Executive Yuan of Taiwan issued a Framework for Taiwan's Sustainable Energy Policy (Executive Yuan 2008). The objective of the policy is to build a win-win-win solution for energy, the environment and the economy, and to construct an energy supply system with clean, stable, and economical efficiency. The principles of the policy are "Two Highs, Two Lows": high efficiency, high value-added, low emissions, and low dependency. High efficiency means improving energy consumption and transformation efficiency. High value-added means increasing the incremental value of energy usage. Low emissions means adopting energy supply methods and consumption practices that ensure low-carbon emissions and low pollution. Low dependency means decreasing Taiwan's dependence on fossil fuels and imported energy. The framework for the policy is "Clean and Reducing" and developing a carbon-free energy and extending the potential usage of renewable energy, so that in 2025 clean energy sources can reach 8% and more of the total energy supply. Therefore, the policy of renewable energy development and promotion is consistent with our expectations.

In order to systematically promote renewable energy, in August of 2002, the Executive Yuan submitted the "Renewable Energy Development Law (REDL)" to the Legislative Yuan, where it remained for 7 years. In June 2009 the Legislative Yuan finally passed the act, which was promulgated by the ROC President in July 2009. The "Renewable Energy Development Law" is comprised of 23 articles and was enacted to promote the utilization of renewable energy sources, increase energy diversification, improve environment quality, energize the industry and drive national sustainable development. (Article 1)

The REDL denotes the total capacity of renewable energy power generation equipment, the target percentage of all types of renewable energy, the power connection and cost allocation of the power industry, the setting of wholesale prices, and the creation of a price-adjusting mechanism. Therefore, the core strategy of REDL is a Feed-in-Tariff system.

First, regarding the promotion of renewable energy, the government shall make steady growth on the installation of renewable energy power generation equipment. The government (MOEA) shall consider the climate and environmental factors, the characteristics of electricity demand and the economic benefits, and the stability of the power supply, while at the same time considering each type of renewable energy development potential, the economic benefits, and key technologies. The government also needs to set promotion goals and the percentage of each category every two years. Taiwan sets the reward capacity for renewable energy power generation equipment as the total capacity between 6,500,000 KW and 10,000,000 KW (Article 4, Article 6).

Second, regarding the executing level of renewable energy, the energy generated by renewable energy power equipment related to power parallel connections, wholesale obligations, and cost sharing shall be interrelated and sold at a wholesale rate by the local power company. The local company shall provide a stable grid and reasonable costs as well. Beyond the existing lines, the cost of installing enhanced power grids is shared by the power company and the operator of the renewable energy power generation equipment. The lines connecting the renewable energy power generation equipment and the power grids shall be built, installed and maintained by the operator of the renewable energy power generation equipment; if necessary, the power company with parallel connection to the power generation equipment shall provide any required assistance; the incurred cost shall be paid by the operator of the renewable energy power generation equipment (Article 8).

Third, regarding the source of the fund, Taiwan's government oversees the renewable energy development fund providing and usage. The power company or the institution operating the self-owned power generation equipment that reaches a certain level of capacity shall pay a certain amount into a fund according to the non-renewable energy portion of the total power generation and the fund is used for renewable energy development. The fund is to be used as follows: to provide subsidies for electricity generation from renewable energy; subsidies for equipment to generate renewable energy, and subsidies for the demonstration of renewable energy and promotion of its use.

The fund-providing subjects includes the power company and the institution that installs the self-owned power generation equipment with certain capacity and has paid the fee to the fund; it may add the paid amount to the selling price of the electricity after approved by the government (Article 7).

In 2011the total budget of renewable energy development fund was NT $2,098,832 thousands. The budget usage of the fund stipulates that the establishment of a data fund on renewable energy basic power generation equipment, renewable energy fund collection and subsides of operators, and renewable energy wholesale price research, et al., were 37,000 thousands; funds for renewable energy usage demonstration and promotion were NT $580,000 thousands, renewable energy subsidies were NT $675,000 thousands, and general administration funds were NT $6,432 thousands (Bureau of Energy, Ministry of Economic Affairs, 2011a).

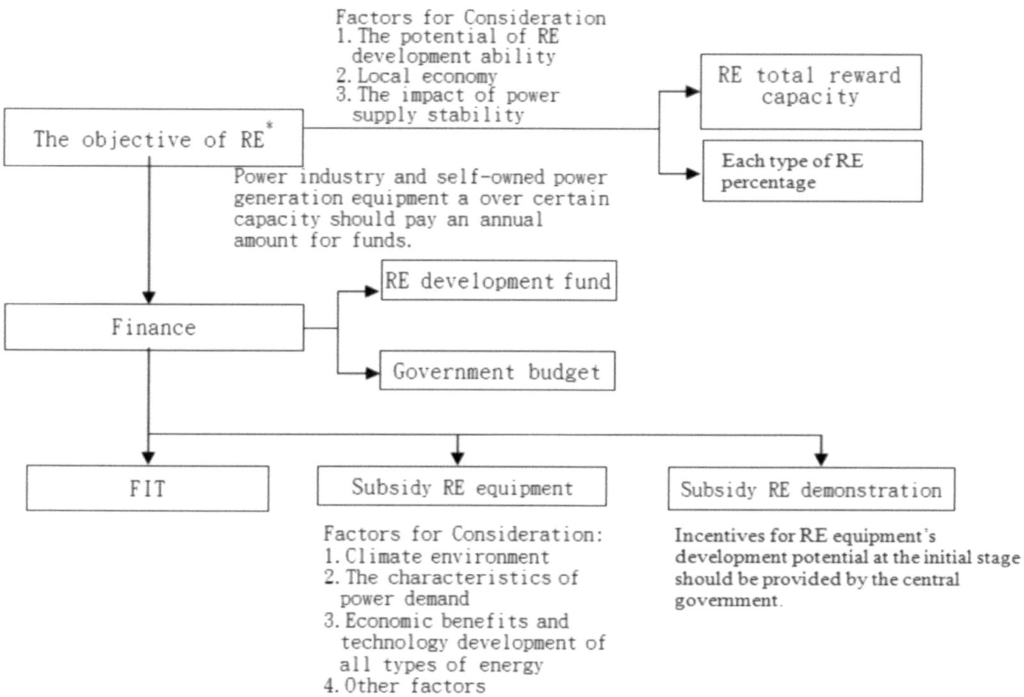

Note: RE= renewable energy
Source: Drawn by the authors based on the Renewable Energy Development Law

Figure 2. Flowchart of the Renewable Energy Development Law

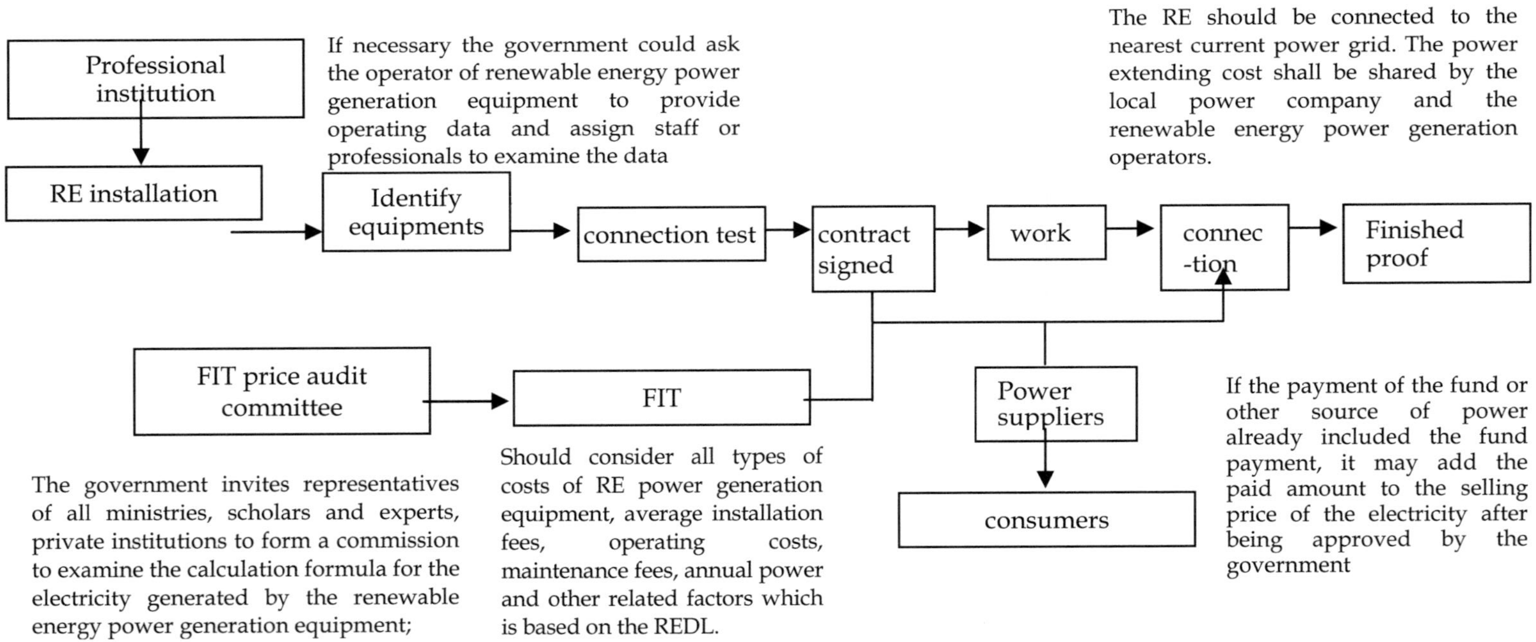

Source: Drawn by the authors based on the Renewable Energy Development Law

Figure 3. Flowchart outlining Taiwan's renewable energy FIT electricity price operation

4. Financial Mechanism of Taiwan's Renewable Energy Feed-in-Tariff

4.1 The Renewable energy wholesale rate examination commission

The government invites representatives of each ministry, and scholars and experts from private institutions to form a commission to examine wholesale rates and the calculation formula for the electricity generated by the renewable energy power generation equipment. If necessary, it shall follow the Administrative Procedure Act to hold hearings and make public announcements; it shall review and revise the rate and the calculation formula every year with respect to advances in power-generating technology, cost variation, progress in reaching goals, and other relevant factors for each category of renewable energy. The above formula for calculating rates is determined by the government, taking into account average installation costs, maximum operating life, annual power generation capacity and other relevant factors of power generation equipment for each category of renewable energy on an individual basis. To encourage and promote pollution-free green energy and increase investment in renewable energy, the wholesale purchase rate shall not be lower than the average cost of the generation of domestic fossil fuels (Article 9).

4.2 Formula for the wholesale rates of renewable energy

According to an announcement made by the Bureau of Energy, Ministry of Economic Affairs, the formula for the wholesale rates of renewable energy is based on all kinds of factors, including installation costs, operating years, maintenance costs, annual power generation capacity, capital cost rate, and reasonable profit rates. The rates are set by the nature of the individual sources such as wind power, river-type hydraulic, geothermal energy, biomass energy, waste, and solar energy, respectively. The rates are effective from January 1st 2011 to December 31st 2011. Starting from 2011 any electricity from new renewable energy power generators will be purchased at this rate for 20 years. In 2011 the reasonable profit rate is 5.25%. Table 9 states the formula for the rates renewable energy.

$$\mathrm{WholsaleRate} = \frac{\mathrm{BICost} \times \mathrm{CRRate} + \mathrm{AOCost}}{\mathrm{AESold}}$$

$$\mathrm{CRRate} = \frac{\mathrm{ACDRate} \times (1 + \mathrm{ACDRate})^{\mathrm{WholesalePeriod}}}{(1 + \mathrm{ACDRate})^{\mathrm{WholesalePeriod}} - 1}$$

$$AOCost = BICost \times \frac{AOCost}{BICost}$$

Note: BICost = Beginning Installation Cost , CRRate = Capital Revert Rate,
AOC = Annual Operating Cost , AESold = Annual Electricity Sold,
ACDRate = Average Capital Discount Rate
Source: Bureau of Energy, Ministry of Economic Affairs, (2011).

Table 9. Formula for renewable energy wholesale rates

4.3 Wholesale rates

The present renewable energy strategies in Taiwan include those for the short run and the long run. The short-term strategy is to prioritize land-based wind power electricity, which has a relatively mature technology and higher economic benefits. The long-term strategy is to encourage the development of offshore wind power electricity. "The wind power generation equipment" can be classified as an offshore wind power system and a land-based wind power system. The offshore wind power rate is 5.5626 NT$ per KWh. Land-based wind power can be divided into two levels: 1 KW to 10 KW and 10 KW and up and the wholesale rates are NT $7.35/KWh and NT $2.61/KWh, respectively.

Category of Renewable Energy	Type	Capacity Level	Wholesale Rate (NT dollar/KWh)
Wind Power	Land-based	1 KW and up 10 KW less	7.3562
		10Kw and more*	2.6138
	Offshore	Indifference	5.5626
River-Type Hydraulic,	Indifference	Indifference	2.1821
Geothermal	Indifference	Indifference	4.8039
Biomass	Indifference	Indifference	2.1821
Waste	Indifference	Indifference	2.6875
Other	Indifference	Indifference	2.1821

Source: Bureau of Energy, Ministry of Economic Affairs, (2011).

Table 10. 2011 Renewable Energy Power (except solar power) Wholesale Rate

The cost of solar photovoltaic electricity generation is much higher than the cost of other sources of renewable energy. Solar photovoltaic generation equipment can be divided into rooftop and ground-mounted. Considering the utility of public land as well as the limited resources of Taiwan, the government has prioritized the development of rooftop solar photovoltaic generation rather than ground-mounted equipment. Residents are encouraged to establish a solar photovoltaic system that can generate from 1 KW to 10 KW of electricity. Rooftop models can be set at 4 different levels: 1 KWto 10 KW, 10 KWto 100 KW, 100 KWto 500 KW, and 500 KW and up. The wholesale rates are NT $ 10.3185/ KWh, NT $9.1799/ KWh, NT $8.8241/ KWh, and NT $ 7.3297/KWh, respectively. Table 9 indicates the 2011 wholesale rates for solar photovoltaic generation equipment.

4.4 Solar photovoltaic bidding mechanism

In order to encourage residents to implement solar photovoltaic generating systems, in 2011the house owners could install a rooftop solar photovoltaic system that generates from 1 KW to 10KW with the wholesale rates on a first come, first served basis at the completion rate and no need to go through the bidding process. Those interested in other types of solar photovoltaic systems are required to go through the bidding process.

Category of Renewable Energy	Type	Capacity Level	Maximum Rate (NT $/KWh)
Solar Photovoltaic generation	Rooftop	1 KW to 10 KW	10.3185
		10 KW to 100 KW	9.1799
		10 KW to 500 KW	8.8241
		500 KW and up	7.9701
	Ground-Mounted	1 KW and up	7.3297

Source: Bureau of Energy, Ministry of Economic Affairs, (2011).

Table 11. 2011 Solar Wholesale Rates for Photovoltaic Generation Equipment

The solar photovoltaic bidding mechanism uses a discount rate in quoting prices. The highest discount rate gets the bid first. The wholesale rate is equal to the completion publicized rate (1-discount rate). Because of limited land availability, the government does not encourage ground-mounted solar photovoltaic systems. If the ground-mounted type of solar photovoltaic generation has the same discount rate as the rooftop version, then the latter is preferred to the ground-mounted type. A rooftop type and a lower capacity are prioritized.. The capacity of each application is limited to the range 1 KW to 2,000 KW.

When applicants make a bid, they must make a deposit. The deposit is based on 1000 times the capacity level, and the deposit should be between NT$ 10,000 and NT $1,000,000. Table 12 denotes the upper limit of the bidding rates and the first period of the 2011 bidding rates for solar photovoltaic generators.

Unit: KW

Period		Rooftop	Ground Mount
First Period		12,000	3,000
	Period Total Upper Limit		15,000
Second Period	First phase	5,000	1,000
	Second phase	5,000	1,000
	Third phase	7,600	1,000
	Period Total Upper Limit		17,600

Source: MOEA (2011) Solar Photovoltaic Generator operating menu modified by the authors.

Table 12. 2011 Upper Limit Bidding Rates

The discount rate of the first period must be greater than 0.00%, but the second period shall be no less than the same level of the first period's average bidding rate. Furthermore, the same period is divided into different bidding phases which keep increasing the average discount rate so that the bidding process requires investors to install renewable energy equipment as soon as possible and to participate in the bidding process. Table 13 indicates the lowest discount rates for all types of solar photovoltaic generators at all levels.

Category of Renewable Energy	Type	Capacity Level	The Same Level of Average Bid Winning Rate
Solar Photovoltaic	Rooftop	1KW to 10 KW	1.24%
		10 KW to 100 KW	2.64%
		10 KW to 500 KW	3.19%
		500 KW and up	0.00%
	Ground-Mounted	1 KW and up	0.31%

Source: Bureau of Energy, Ministry of Economic Affairs (2011b) modified by the authors

Table 13. The First Period of 2011 Discount Rates for Solar Photovoltaic Generators

Table 14 lists the 2011 bidding results for solar photovoltaic generators. According to the table, for the second period the number of applications and winning bids as well as bidding capacity all dramatically increased from the first period. Since the policy encouraged rooftop systems and restricted the ground-mounted version, the number of rooftop systems is far greater than ground-mounted systems.

4.5 The Structure of Renewable Energy Capacity

Currently, the type of renewable energy in Taiwan with the largest capacity is a river-type hydraulic generator. Taipower and private companies generate 197.5 MW. Biomass ranked number 2 and generated 80.94 MW. This is followed by wind power at 52.93 MW and solar photovoltaic power at 4.48 MW, respectively. The solar photovoltaic system generated 1.1 MW before the renewable energy development law was passed. After the law was passed the capacity was increased to a range from 2.08 MW to 3.13 MW. In June 2011 the total capacity of renewable energy was 336 MW which comprises 6.9% of total power capacity. Table 15 illustrates the capacity generated from each type of renewable energy source.

5. Conclusion and Recommendations for Taiwan's Feed-in Tariff

Designing a reasonable wholesale rate is the most important issue with respect to a renewable energy feed-in tariff system. One must consider average installation cost, operating life, maintenance cost, annual power generation capacity and relevant factors for different types of power generation equipment separately and set wholesale rates for each category of the renewable energy so that the price not only can ensure an optimal developing opportunity for each type of renewable energy,; one also can reduce the incentive of higher profitable technologies and avoid shifting a heavy cost burden to consumers (Chou, Lin and Chen 2010). To arrive at a reasonable wholesale rate, one also needs to consider size and location. Different locations and sizes generate different electricity costs so and command different wholesale rates. The larger the size is, the more economical the scale. Generating equipment of a larger size has lower average electricity production costs. Therefore, in order to ensure adequate profits, renewable energy generating

Unit: KW

	Type	Number of Applications	Number of Bids	Bidding Capacity	Average Discount Rate
The First Period	Rooftop	126	123	12,173.123	2.62%
	Ground Mounted	2	2	1,379.400	0.31%
	Sum	**128**	**125**	**13,552.523**	-
The Second Period Phase 1	Rooftop	43	40	2,583.181	2.95%
	Ground Mounted	1	1	248.640	0.31%
	Sum	44	41	2,831.821	-
Phase 2	Rooftop	48	38	4,840.830	3.12%
	Ground Mounted	1	1	110.400	0.31%
	Sum	49	39	4,951.230	-
Phase 3	Rooftop	-	87	7,235.874	3.37%
	Ground Mounted	-	0	0	0.00%
	Sum	100	87	7,235.874	-
The Second Period	**Total Sum**	**193**	**167**	**15,018.925**	-

Source: Bureau of Energy, Ministry of Economic Affairs (2011b) modified by the authors

Table 14. 2011 Bidding Results for Solar Photovoltaic Generating Equipment

Item	Wind Power		Biomass			Total	Percentage of Total Electricity Power
	TP	Private	Municipal Solid Waste	Agriculture & Industry Solid Waste	Biogas		
Capacity (Million W)	28.88	24.05	62.25	16.75	1.94	336	6.91%
Sum	52.93		80.94				
Item	River-Type Hydraulic		Solar Photovoltaic				
	TP	Private	Finished Before Law*	Finished after Law	TP Self-owned		
Capacity (Million W)	193.6	3.9	1.1	3.13	0.25		
Sum	197.5		4.48				

Note: 2011/6/30, Law* Refer to the Renewable Energy Development Law

Source: Industrial Technology Research Institute Renewable Energy(2011).

Table 15. Capacity of Each Type of Renewable Energy

equipment of a larger capacity should be paired with a diminishing marginal rate. One location might have greater wind power; therefore, when a wind power generator is land-based or offshore, different costs different capacities of electricity ensue. Usually, the government assigns a higher wholesale rate to a priority location to increase incentives to produce more renewable energy.

The present renewable energy strategies in Taiwan include those for the short run and the long run. The short-term strategy prioritizes land-based wind power electricity, because the technology is relatively mature and the economic benefits are higher. The long-term strategy is to encourage the development of offshore wind power electricity. In terms of solar photovoltaic sources, when one considers the total utilization of land and the limited land resources of the nation, Taiwan prioritizes rooftop solar photovoltaic systems and discourages the implementation of ground-mounted systems. A resident owner who installs a rooftop solar photovoltaic system with a capacity from 1 KW to 10 KW qualifies for a priority subsidy and the highest wholesale rate.

The "wholesale rate guarantee period" is another important design feature. The longer the guarantee period is, the lower the investment risk is. The length of wholesale rate is determined by the time of the return on investment, the operating life of renewable energy equipment,, equipment renewal speed, and loan provisions. In Taiwan there is a guarantee period of twenty years for those who are already in the system and for whom the risk is low. The FIT price is reexamined every year; however, for those who have not entered the system the risk is high. Therefore, Taiwan publically discloses the FIT electricity price every year so that the public can examine and discuss the rate.

In order to create an environment capable of developing competitive renewable energy sources, the government provides incentive mechanisms such as capital subsidies, investment tax credits, et al. (REN21, 2011). Capital subsidies refer to government subsidies awarded to those who install renewable energy systems to reduce their capital burden and increase investment opportunities. Investment tax credits means the government permits investors to deduct their investment in renewable energy equipment from their tax liabilities to lower the investors' tax burden. Furthermore, an energy tax is another means for supporting such measures. An energy tax is imposed on those using traditional fossil fuels that increase greenhouse gases. The tax may force investors to internalize the social costs which lower the cost of the renewable energy and allow consumers to choose from different sources of renewable energy. Taiwan's government has also set regulations for renewable energy subsidies regulation and implemented a renewable energy tax credit act. However, the energy tax in Taiwan has not been passed; the bill still needs to be negotiated. In the future, the burden of energy costs need to be fairly distributed and the rights of socially vulnerable groups should be protected.

By all accounts, renewable energy FIT can bring about environmental, economic, and social benefits, and can promote the renewable-energy industry as well. By applying REDL, Taiwan is going to initiate a new opportunity of renewable energy investment and is moving toward the creation of a low-carbon society.

6. References

[1] Bureau of Energy, Ministry of Economic Affairs (2010), Energy Statistics Handbook 2010.

[2] ______, (2011), Announcement.

[3] ______ (2011a), 2011 Republic of China Center Authority Budget. http://www.moeaboe.gov.tw/opengovinfo/budgetlist.aspx

[4] ______ (2011b), http://www.moeaboe.gov.tw/board/boarddetail.aspx?serno=01238

[5] Chou, L. F., L. F. Lin and S. M. Chen, 2010.11,"The Price Mechanism and Tax Incentive of Renewable Energy Feed-in-Tariff", Tax Research,Vol.42,No.6, pp.61-78.

[6] Directorate General of Budget, Accounting and Statistics (DGBAS) of Executive Yuan, http://www.dgbas.gov.tw/ct.asp?xItem=393&CtNode=2850&mp=1

[7] Executive Yuan (2008), Framework of Taiwan's Sustainable Energy Policy.

[8] Industrial Technology Research Institute Renewable Energy (2011), http://www.re.org.tw/re2/impetus.htm.

[9] Klein, A., E. Merkel, B. Pfluger, A. Held, and M. Ragwitz, (2008), Evaluation of Different Feed-in Tariff Design Options - Best Practice Paper for the International Feed-in Cooperation.

[10] Mendonça, M., D. Jacob, and B. Sovacool, (2010), Powering the Green Economy: The feed-in Tariff Handbook, RFF Press.

[11] REN21, (2011), Renewables 2011: Global Status Report, http://www.ren21.net/

[12] The Policy Action on Climate Toolkit PACT (2011), http://www.futurepolicy.org/renewableenergy.html

3

Assessment of the Decoupling of GHGs and Electricity Costs Through the Development of Low-Carbon Energy Technology in Taiwan

Chien-Ming Lee and Heng-Chi Liu
Institute of Natural Resource Management, National Taipei University
Taiwan Research Institute
Taiwan

1. Introduction

Since the 1990s, global warming together with the abatement of greenhouse gases (GHG) has emerged as a key issue in the world. Achieving a 450mmp GHG concentration in the atmosphere and a control temperature of less than 20C relative to pre-industrialized conditions in the world in 2100 have been designated as long-term goals. The International Energy Agency (IEA, 2008) indicated that low-carbon energy technologies (including renewable energy and biofuels, nuclear energy, natural gas, et al.) are priority policies and measures to response to global warming and reach GHG mitigation targets, where energy efficiency and renewable energy account for about 78% of the reduction of GHG emissions (see Figure 1).

The energy sector accounted for more than 66% of GHG emissions in 2008; it is the biggest GHG emission sector in Taiwan (see Figure 2). Thus, how to reduce CO2 emissions from power generation has become the most important strategy in response to global warming in Taiwan. Therefore, Taiwan's government passed "The Sustainable Energy Policy Guidance "in 2008; it also established a low-carbon energy target in 2020 as well, i.e., it has deployed low-carbon energy , with a goal of up to 55% (renewable energy no less than 8%, natural gas must more than 25%) in power generation in 2025. Figure 3 indicates the 40.6% low-carbon energy rate in 2008; in other words, a huge gap (i.e., a reduction of about 15%) needs to be closed in the coming decade.

In addition, under "The Sustainable Energy Policy Guidance," Taiwan's government has committed itself to reducing CO2 emissions to the 2008 level (about 294 MtCO2) by 2016-2020, and to the 2000 level (about 221 MtCO2) by 2025. However, due to the lack of previous CO2 emission reduction assessments, it is not clear whether the ambitious GHG target can be achieved by 2025.Besides, how will electricity costs be impacted? This is a significant concern of the public. The purpose of this paper is to assess the effect of GDP decoupling with GHG emissions and the impact of the cost of electricity by developing low-carbon energy technology in Taiwan. Implications for the government with respect to policy implications will also be provided.

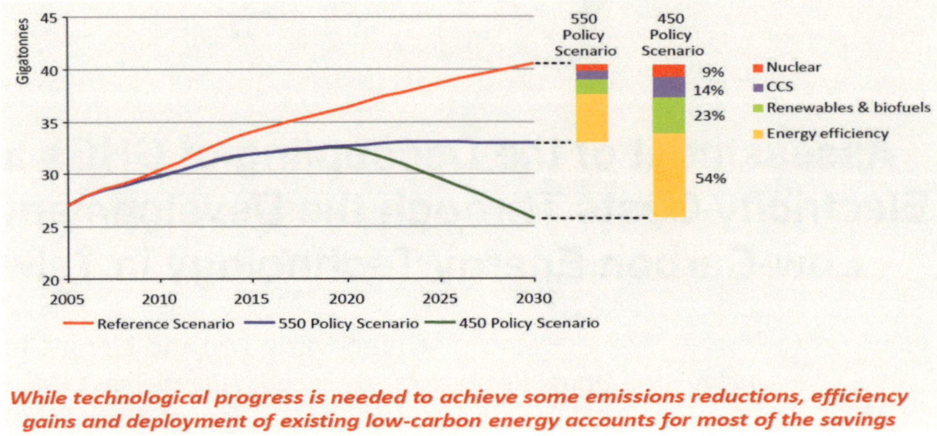

Source: IEA (2008), World Energy Outlook 2008.

Figure 1. GHG abatement strategy in various climate scenarios

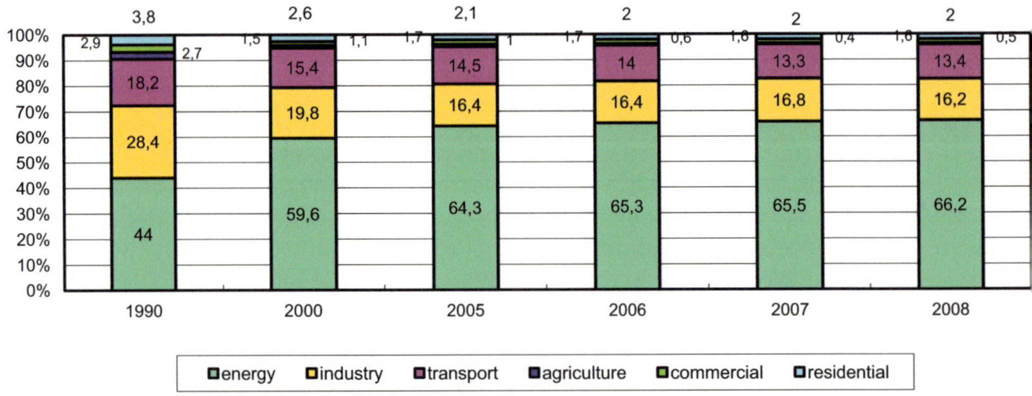

Source: Bureau of Energy (2009), Trend of CO2 emission rates from fuel combustion in various sectors in Taiwan.

Figure 2. Trend of CO2 emission share in various sector in Taiwan

2. Methodology

Equation 1 illustrates CO_2 intensity (CO_2 / GDP) can be broken down into two parts, CO_2 emission per Energy (CO_2 / E) and energy intensity (E / GDP) respectively, where CO_2 / GDP is a decoupling indicator; i.e., if CO_2 / GDP is reduced, this will result in GDP decoupling with CO_2 emissions. CO_2 / E represents a degree of clean energy (or low-carbon energy) in power generation; in other words, CO_2 / E will be reduced if the clean energy share of power generation increases. E / GDP is the inverse of energy efficiency, meaning E / GDP will decrease when energy efficiency increases.

Equation 2 illustrates how CO2 intensity ($CO_2^{total} / GDP^{total}$) nationwide can be divided into four parts: (1) CO2 emission per Energy ($CO_2^{energy\ sector} / E^{energy\ sector}$) in the energy sector; (2)

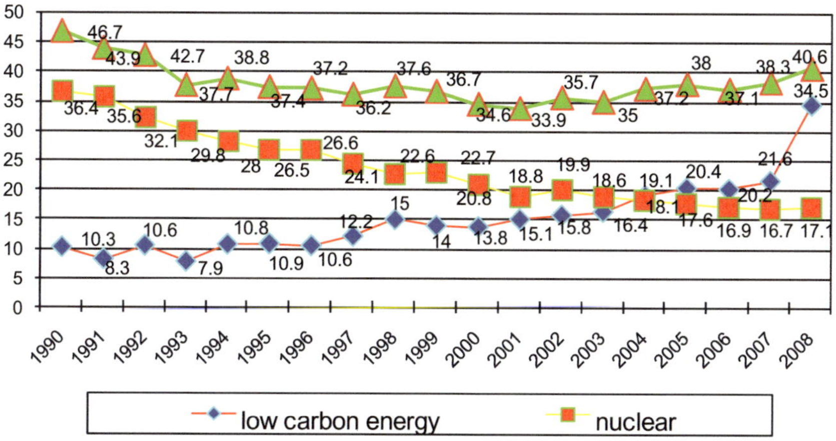

Source: Bureau of Energy (2009), Trend of CO2 emission rates from fuel combustion in various sectors in Taiwan.

Figure 3. Trend of low-carbon energy rates in Taiwan

energy consumption share $(E^{energy\ Sector}/E^{total})$ in the energy sector; (3) inverse CO_2 emission share(CO_2^{total} / $CO_2^{energy\ sector}$) in the energy sector, and (4) energy intensity (E^{total}/GDP^{total}) nationally.

As all of the penal data are time series, "unit root" and "co-integration" tests, these must be engaged in before regression can be run.[1] In addition, this study adopts a mean absolute percentage error (MAPE) criterion to make sure the regression equations can be used to predict a future time path.[2] (See Figure 4)

$$\frac{CO_2^{total}}{GDP^{total}} = \frac{CO_2^{total}}{E^{total}} \times \frac{E^{total}}{GDP^{total}} \tag{1}$$

$$\frac{CO_2^{total}}{GDP^{total}} = \frac{CO_2^{enrergy\ sector}}{E^{energy\ sector}} \times \frac{E^{energy\ sector}}{E^{total}} \times \frac{CO_2^{total}}{CO_2^{emergy\ sector}} \times \frac{E^{total}}{GDP^{total}} \tag{2}$$

[1] This indicates that a stable relationship exists among dependent and independent variables.
[2] A MAPE of less than 10% means it is highly accurate; one greater than 50% is not accurate.(Lewis,1982)

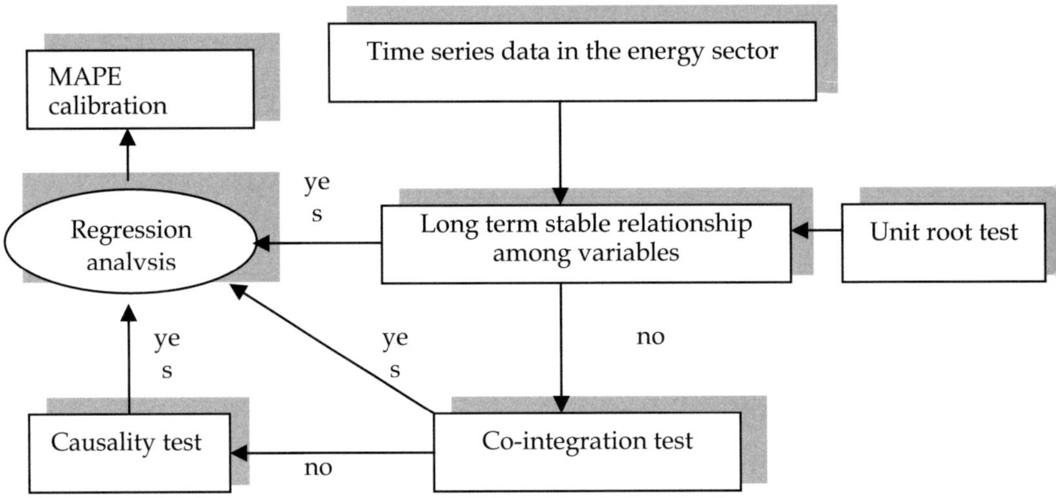

Figure 4. Flowchart illustrating the econometric test process

3. Scenario design and Incorporating the learning effect

To simplify the study, a scenario has been designed as follows:
1 Allow energy efficiency (E/GDP) to increase 2% annually.
2 Set two low-carbon energy power generation rate scenarios at 55%, and 60%, respectively.
Experience studies have demonstrated that there is a correlation between the cost of manufacturing an item and the cumulative quantity of the item produced (Colpier and Cornland, 2002; Hamon, 2000; Neij, 1999).This relationship can be illustrated by an experience curve, which shows that the cost of a product decreases by a certain percentage every time the total quantity manufactured (total experiences) doubles. The experience curve is often expressed as a power function. (See following equation.)

$$C_q = C_0 q^{-b} \tag{3}$$

Where C_q is the cost per unit q, C_0 is the cost for the first unit, q is the cumulative production (experience curve time) and b is a so-call experience index. The value 2^{-b} is called the progress ratio (PR). If an experience curve shows a progress ratio of 85%, it means that cost declines by 15% (learning rate) for each doubling of cumulative production. The reduction of the average cost of power generation is the result of the learning effect (See Appendix). This is derived from the progress ratio estimation of the average cost of power generation average cost in Taiwan (See Table 1).

year	PR (%)	leArning rate (%)
2009	88.2	11.8
2010	86.7	13.3
2015	80.5	19.5
2020	76.2	23.8
2025	72.5	27.5

Table 1. Progress ratio estimation of power generation costs in Taiwan

4. Results

4.1 Assessment of GDP decoupling with CO2 emission

To simplify the study, we let energy efficiency (E/GDP) increase 2% annually and set two scenarios for low-carbon energy power generation rates of 55%, and 60%, respectively. Figure 5 shows a typical business scenario: CO_2 intensity is 22.3 tCO_2/MNT$ by 2025; however CO2 intensity will be sharply reduced to 10.35 tCO_2/MNT$ in the first scenario of a 55% reduction by 2025. This can be further decreased to 9.06 tCO_2/MNT$ in the second scenario of 60% reduction by 2025. From the above results, it can be easily understood that if the Taiwanese government implements low-carbon energy technology, GDP decoupling from CO2 emissions will be achieved in the future.

unit: tCO_2/MNT$

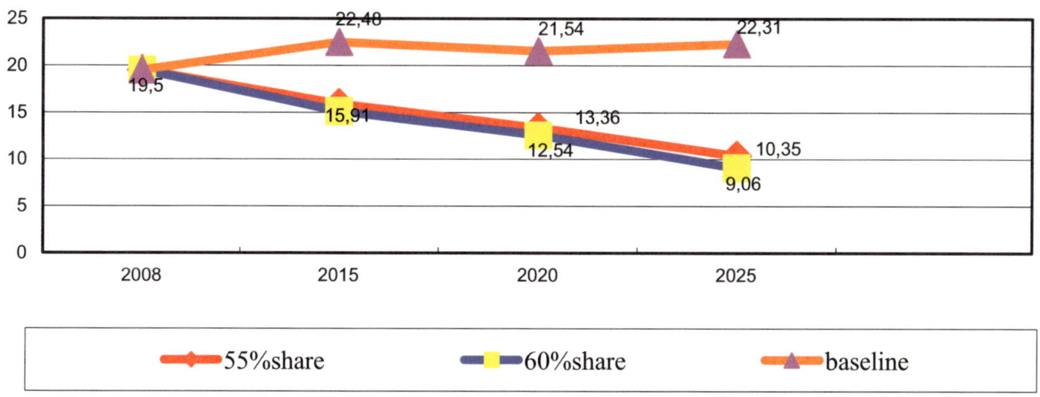

Figure 5. Assessment of GDP decoupling from CO2 at various low-carbon energy rates

4.2 Assessment of electricity costs

Due to the fact that the cost of low-carbon energy is higher than carbon-intense fuels (such as coal), renewable energy sources will increase the share of low-carbon energy sources in power generation. This must then increase electricity costs as well. As indicated in Figure 6, electricity costs will significantly increase to 7.25 NT$/kWh in the 55% scenario, and 7.65 NT$/kWh in the 60% scenario.

5. Conclusion

Under the Framework on Sustainable Energy Development Policies developed by Taiwan's government, the low-carbon technology development target (i.e., not less than 55%) is to be reached as a response to GHG mitigation by 2025. The purpose of this paper is to assess the effect of GHG decoupling and electricity costs by the development of low-carbon energy technology in Taiwan. Results indicate the following: CO_2 is decoupled with economic growth when electricity generation rates of low-carbon energy go up. This can be seen from the following: (1) CO_2 intensity decreases from 22.31 tCO_2/MNT$ (in 2025) to 10.35tCO_2/MNT$ (in 2025) if the electricity generation rate of low-carbon energy reaches

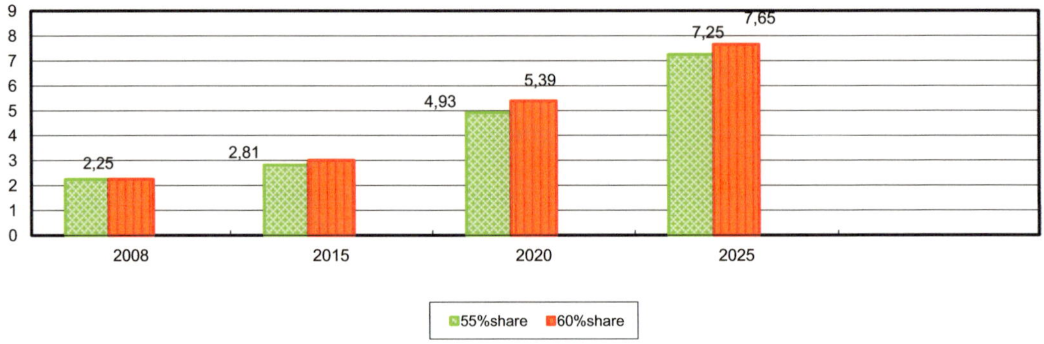

unit: NT$/kWh

Figure 6. Electricity cost in various low carbon energy share assessment

55%; (2) CO_2 intensity decreases from 22.31tCO_2/MNT$ (in 2025) to 9.06 tCO_2/MNT$(in 2025) if the electricity generation rate of low-carbon energy reaches 60%. However, electricity costs also occur as the electricity generation rate of low-carbon energy increases; i.e., (3) the cost of electricity will increase from 2.25 NT$/kWh (in 2008) to 7.25 NT$/kWh (in 2025) if the electricity generation rate of low-carbon energy reaches 55%; (4) the cost of electricity will increase from 2.25 NT$/kWh (in 2008) to 7.65 NT$/kWh (in 2025) if the electricity generation rate of low-carbon energy reaches 60%.

6. References

[1] Bureau of Energy (2009), Trend of CO2 emissionS from fuel combustion in various sectors in Taiwan.

[2] Chen, T.Y., Yu, O.S., Hsu, G.J., Hsu, F.M., and W.N., Sung, (2009), "Renewable energy technologyportfolio planning with scenario analysis: A case study for Taiwan", Energy Policy, 37(8): p.2900-2906.

[3] Colpier, C. U. and D. Cornland (2002), "The economics of the combined cycle gas turbine-an experience curve analysis", Energy Policy, Vol. 30, p.309-316.

[4] Hamon, C. (2000), "Experience curves of photovoltaic technology", IIASA Interim ReportIR-00-014, Austria; International Institute for Applied Systems Analysis.

[5] IEA (2008), World Energy Outlook 2008.

[6] Neij, L. (1999), "Cost dynamics of wind power", Energy, Vol. 24, p.375-389.

7. Appendix

$\ln \frac{CO_2{}^i}{E^i} = c_0 + c_1 \ln S_1 + c_2; \ln S_2 + c_3 \ln S_3 + c_4 \ln S_4 + c_5 \ln S_5 + c_6 \ln S_6 + c_7 \ln S_7$		
variables	Statistics coefficient	T value
constant	1.2267	0.7805
$\ln(S_1)$	-0.0180	-0.3077
$\ln(S_2)$	0.0843	1.3490
$\ln(S_3)$	0.0179	0.0636
$\ln(S_4)$	0.1939	1.2938
$\ln(S_5)$	-0.0908	-1.1346
$\ln(S_6)$	-0.0040	-0.5035
$\ln(S_7)$	-0.3355	-1.2657
R-squared	0.9775	

Table A1. CO_2/E regression equation

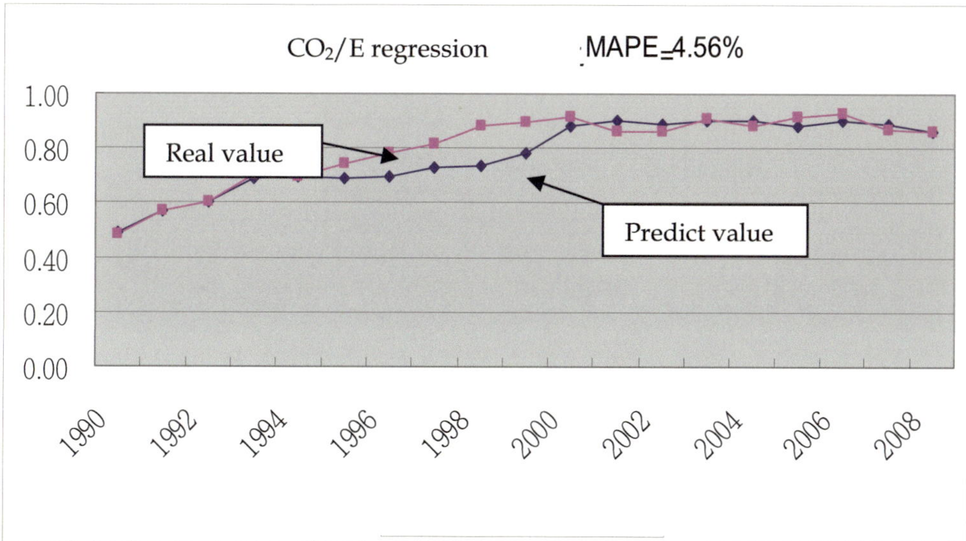

Figure A1. MAPE calibration of CO_2/E

$AC_{1_t}^{e} = c_0 + c_2; S_2 + c_3 S_3 + c_4 S_4$		
variables	Statistics coefficient	T value
Constant	3.436733	5.422285
Oil share(s_2)	-0.0302	-1.65073
Coal share(s_3)	-0.04144	-2.97093
CHP share(s_4)	-0.07528	-3.48088
R-squared	0.87763	
$AC_{2_t}^{e} = c_0 + c_1 S_1 + c_5 S_5 + c_6 S_6 + c_7 S_7$		
Constant	1.160219	1.514886
Natural gas share(s_1)	0.031062	1.511986
Hydro share(s_5)	-0.0039	-0.1836
Renewable energy share(s_6)	3.043012	5.705584
Nuclear share(s_7)	0.001841	0.093962
R-squared	0.962363	

Table A2. Average electricity cost (AC) regression equation

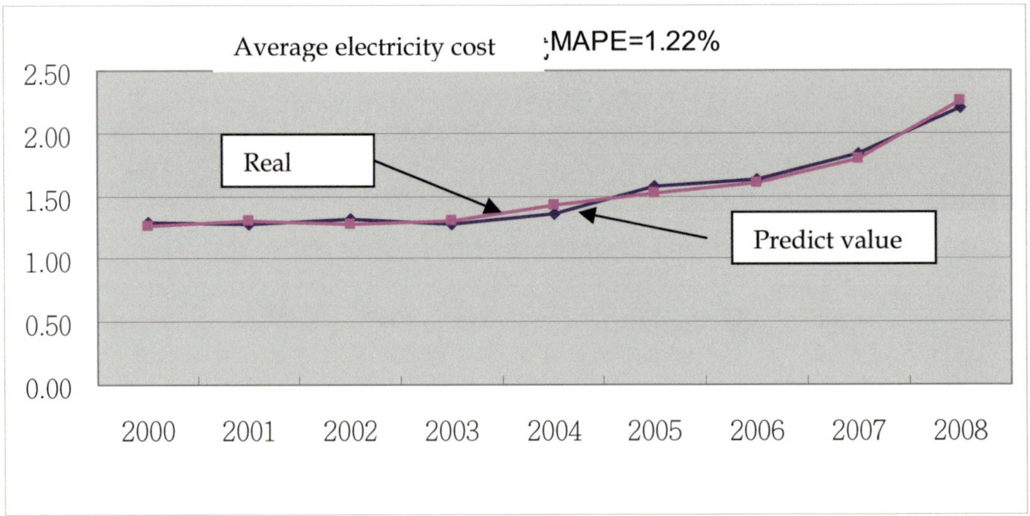

Figure A2. MAPE calibration of average cost

$\ln C_t = \ln C_0 - \zeta \ln Q_t + t + \varepsilon_t$		
variables	Statistics coefficient	T value
constant	14.228	17.538
$\ln Q_t$	-0.178	-2.706
t	0.136	8.335
R-squared	0.980	

Table A3. Learning curve estimation of power generation

4

Estimation of Taiwan's CO_2 Emissions Related to Fossil Fuel Combustion – A Sectoral Approach

Shinemay Chen, Der-Cherng Lo and Huai Hsuan Yu
Department of Public Finance, National Cheng-Chi University
Taiwan

1. Introduction

In terms of annual carbon dioxide (CO_2) emissions, Taiwan emitted 293.66 million metric tons of CO_2 in 2007 and the volume was down to 279.14 million metric tons in 2009. However, from 2007 to 2009, Taiwan's CO_2 emission ranking rose from the 22nd to the 21st largest emitter in the world. International comparisons of total CO_2 emissions are shown in Table 1. After the Kyoto Protocol entered into force in 2005, the Taiwanese government convened its second National Energy Conference.[3] The Taiwan Environmental Protection Administration (EPA), designated as the leading government agency in greenhouse gas policy, submitted its Greenhouse Gas Reduction Bill to the legislature in 2006. Unfortunately, the Greenhouse Gas Reduction Bill was not passed.

After President Ma Ying-jeou took office in 2008, he announced his target of stabilizing Taiwan's GHG emissions at 2008 levels by 2020. Furthermore, the Committee of Carbon Reduction of the Executive Yuan has proposed a national target for reducing carbon dioxide in fuel emissions, dropping to 2005 levels by 2020 and to 2000 levels by 2025. The EPA resubmitted the Greenhouse Gas Reduction Bill to the legislature in 2008. It is still being considered, but if it passes, the bill would authorize the EPA to regulate GHGs with a cap-and-trade scheme and sectoral emission performance standards. That is, the government of Taiwan is considering setting up a carbon trading exchange.

Accordingly, the understanding of the historical allocation of the carbon dioxide emission across sectors and industries becomes very important. This information will allow the government to evaluate the potential trading volume of a future domestic carbon market. To get a grip on the issue of potential trading volume, we start from estimating Taiwan's CO_2 emission levels. Since the largest source of CO_2 emissions is from the oxidation of carbon

---

[3] As a response to the Kyoto Protocol, the government convened the first National Energy Conference in 1998.

Units: Million Metric Tons

Ranking	Country	Year 2007	Ranking	Country	Year 2009
1	China	6256.704	1	China	7706.826
2	United States	6018.131	2	United States	5424.53
3	Russia	1627.203	3	India	1591.126
4	India	1368.383	4	Russia	1556.661
5	Japan	1254.438	5	Japan	1097.965
6	Germany	827.2343	6	Germany	765.5618
7	Canada	610.0027	7	Canada	540.9669
8	United Kingdom	569.8945	8	Iran	528.6026
9	Korea, South	503.0997	9	Korea, South	528.1344
10	Iran	489.3254	10	United Kingdom	519.944
11	South Africa	463.7263	11	South Africa	451.2196
12	Italy	459.5288	12	Mexico	443.6122
13	Mexico	444.2595	13	Saudi Arabia	438.2468
14	France	423.0563	14	Brazil	425.1693
15	Australia	410.3513	15	Australia	417.6815
16	Brazil	400.4417	16	Indonesia	414.9409
17	Saudi Arabia	396.4678	17	Italy	407.866
18	Indonesia	390.2196	18	France	396.6518
19	Spain	387.9257	19	Spain	329.8573
20	Ukraine	354.0988	20	Poland	285.7852
21	Poland	295.9488	21	Taiwan	279.1429
22	Taiwan	293.6621	22	Thailand	254.8797
23	Turkey	280.1906	23	Turkey	253.0567
24	Netherlands	258.1038	24	Ukraine	252.4726
25	Thailand	247.3535	25	Netherlands	248.9068

Source: International Energy Statistics, U.S. Energy Information Administration, http://www.eia.gov/cfapps/ipdbproject/IEDIndex3.cfm?tid=90&pid=44&aid=8.

Table 1. Total Carbon Dioxide Emissions from the Consumption of Energy

when fossil fuels are burned,[4] IPCC makes sectoral emissions estimates based on fuel consumption. In Taiwan's recent inventories of greenhouse gas emissions, estimates of CO_2 emissions carried out by the Bureau of Energy were calculated in accordance with IPCC methodology. However, the estimation breaks down the CO_2 emissions by sector (energy

[4] Fossil fuel combustion accounts for 70-90 percent of total anthropogenic CO_2 emissions. (IPCC, 1997a)

sector, industrial sector, transportation sector, agricultural sector, service sector, and residential sector).

Attempting to delineate the CO_2 emissions of Taiwan, we further estimate Taiwan's fossil fuel CO_2 emissions by subsector/industry.

2. Research Methodology

In this section, the methodology for estimating CO_2 emissions from energy is discussed. Following the IPCC Guidelines for National Greenhouse Gas Inventories,[5] we estimate 2005-2010 CO_2 emissions based on fuel consumption data by sub-sectoral activity. Individual subsector's fuel consumption data were obtained from the Energy Balance Sheet compiled by the Bureau of Energy.[6] Table 2 gives a list of subsectors and industries included in the sample.

Therefore, CO_2 emissions from fuel combustion by fuel type j for each subsector (industry) i in year t is calculated as follows:[7]

$$\boldsymbol{CO_2\text{emission}_{i,t}} = \sum_j [\boldsymbol{carbon\ content\ for\ fuel\ j - carbon\ stored\ for\ fuel\ j}] \\ * \boldsymbol{fraction\ of\ carbon\ oxidized\ for\ fuel\ j} * \frac{\mathbf{44}}{\mathbf{12}} \\ + \boldsymbol{CO_2}\ \textbf{emissions from electricity consumption for each subsector}\ \boldsymbol{i}$$

where

$$\boldsymbol{carbon\ content\ for\ fuel\ j} \\ = \boldsymbol{fuel\ consumption\ expressed\ in\ energy\ units\ (TJ) for\ fuel\ j} \\ * \boldsymbol{carbon\ emission\ factor\ for\ fuel\ j};$$

$$\boldsymbol{carbon\ stored\ for\ fuel\ j} = \\ \boldsymbol{carbon\ content\ for\ fuel\ j} * \boldsymbol{fraction\ of\ carbon\ stored\ for\ fuel\ j};$$

$$\boldsymbol{CO_2}\ \textbf{emissions from electricity consumption for each subsector}\ \boldsymbol{i} \\ = \textbf{total}\ \boldsymbol{CO_2\ emissions\ from\ electricity} \\ * [\boldsymbol{electricity\ consumption\ for\ subsector\ i} \\ /(\boldsymbol{total\ electricity\ consumption})]$$

---

[5] The IPCC Guidelines were first accepted in 1994 and published in 1995. UNFCCC COP3 held in 1997 in Kyoto reaffirmed that the Revised 1996 IPCC Guidelines for National Greenhouse Gas Inventories should be used as "methodologies for estimating anthropogenic emissions by sources and removals by sinks of greenhouse gases" in calculation of legally-binding targets during the first commitment period. http://www.ipcc-nggip.iges.or.jp/public/gl/invs1.html.

[6] http://www.moeaboe.gov.tw/English/Statistics/EnStatistics.aspx.

[7] http://www.ixon.com.tw/

Sector	Subsector or industry	
Energy sector	Coal Mines Coke Ovens Blast Furnaces Oil and Gas Extraction Petroleum Refineries	Electricity Plants Electricity to Pump Up Cogeneration Plants Gas Companies
Industrial sector	Mining and Quarrying Food, Beverage and Tobacco Textile, Wearing Apparel and Accessories Leather and Fur Wood, Bamboo and Furniture Pulp, Paper and Paper Product Printing Basic Industrial Chemicals Petrochemical Materials Chemical Fertilizers Artificial Fibers Resin, Plastics and Rubber Other Chemical Materials Chemical Products	Rubber Products Plastic Products Cement and Cement Products Others (Pottery, China, and etc.) Iron and Steel Non-metal Fabricated Metal Products Machinery and Equipments Electrical and Electronic Machinery Transport Equipments Precision Instruments Miscellaneous Industries Water Supply Construction
Transportation sector	Domestic Air Road	Rail Internal Navigation
Agriculture Sector	Agriculture, Animal Husbandry and Forestry Fishing and Aquaculture	
Service Sector	Wholesale and Retail Hotels and Restaurants Transport Services Storage and Warehousing Communication	Finance, Insurance and Real Estate Business Services Social and Personal Services Public Administration Not Specified Services
Residential Sector		

Table 2. Subsectors or Industries included in the sample

That is, using IPCC methodology as the basis, the estimation process can be divided into six steps that lead to figures for CO_2 emissions from fuel combustion.

1) Obtain the amount of each fuel consumed by each sub-sector.
Since heating value data provided by the Bureau of Energy are in 10^7 kilocalories, we multiply the consumption by 0.04184 to give the amounts of all fuels in terajoules (TJ).

2) Estimate total carbon content in fuels.
Carbon content represents the total amount of carbon that could be emitted if 100 percent were released to the atmosphere. To estimate the carbon content in tons of carbon, we multiply fuel consumption in TJ by the appropriate carbon emission factors (more precisely, the specific carbon content, t C/TJ). This calculation should be done for all fuel types in each sector. The carbon emission factors for each fuel type are shown in Table 3.

3) Estimate the amount of carbon stored in products.
After estimating the total carbon contained in the fuels, the next step is to estimate the amount of carbon from those fuels which are used for non-energy purposes. Some of the fuel supplied to an economy is used as a raw material (or feedstock) for the manufacture of products or in a non-energy use (e.g., bitumen for road construction, lubricants). Therefore, in some cases, the carbon from the fuels is oxidized quickly to CO_2, while in other cases the carbon is stored in the product, sometimes for as long as centuries. The amounts of stored carbon should be deducted from the calculation for total carbon emissions.

Fuel	Carbon Emission Factor (t C/TJ)	Fuel	Carbon Emission Factor (t C/TJ)
Solid Fossil		Liquid Fossil	
Anthracite	26.8	Crude Oil	20.0
Coking coal	25.8	Lubricants	20.0
Lignite	27.6	LPG	17.2
Peat	28.9	Natural Gasoline	17.2
Coke Oven Coke	29.5	Naphtha	20.0
Patent Fuel	25.8	Motor Gasoline	18.9
Coke Oven/Gas Coke	29.5	Aviation Gasoline	18.9
Bituminous Coal- Steam Coal	25.8	Jet Fuel- Kerosene	19.5
Sub-bituminous - Coal	26.2	Jet Fuel- Gasoline	19.5
Blast Furnace Gas	66	Kerosene	19.6
Refinery Gas	18.2	Diesel Oil	20.2
Oxygen Steel Furnace Gas	13	Fuel Oil	21.1
Coke Oven Gas	13.0	Asphalts	22.0
Gaseous Fossil		Solvents	20.0
Natural Gas (dry)	15.3	Petroleum Coke	27.5
		White Spirits	20.0
		Other Petroleum	20.0
		Paraffin Waxes	20.0
		Natural Gas Liquids	17.2

Source: IPCC (1997b). Revised 1966 IPCC Guidelines for National Greenhouse Gas Inventories: Workbook (Volume 2), http://www.ipcc-nggip.iges.or.jp/public/gl/guidelin/ch1wb1.pdf.

Table 3. Carbon Emission Factors

The amount of the carbon stored is obtained by multiplying the carbon content and fraction of carbon stored. Table 4 represents the fraction of carbon stored for different types of fuel. Since in Taiwan, lubricants and bitumen are used as raw materials and in non-energy consumption, the Bureau of Energy gives the figures of these two fuel types as 1.0. Naphtha and LPG are used as raw materials in the industry of Petrochemical Materials; consequently, the Bureau of Energy gives the figures of these two fuel types in Petrochemical Materials as 1.0, while in other industries as 0.

4) Account for carbon oxidized during combustion.
When energy is consumed, not all of the carbon in the fuel oxidizes to CO_2. Incomplete oxidation occurs due to inefficiencies in the combustion process that leave some of the carbon unburned or partly oxidized as soot or ash. The Intergovernmental Panel on Climate Change (IPCC) guidelines for calculating emissions inventories require that an oxidation factor be applied to the carbon content to account for the small portion of the fuel that is not oxidized into CO_2. Table 5 shows the fraction of carbon oxidized. For example, for all oil and oil products, the oxidation factor used is 0.99 (i.e., 99 percent of the carbon in the fuel is eventually oxidized, while 1 percent remains un-oxidized).

Fuel Type	IPCC version	Taiwan's version	
		Petrochemical	others
Naphtha	0.75	1.0	0.0
Lubricants	0.5	1.0	1.0
Bitumen	1.0	1.0	1.0
Coal Oils and Tars (from Coking Coal)	0.75		
Natural Gas	0.33		
Gas/Diesel Oil	0.50		
LPG	0.80	1.0	0.0
Ethane	0.80		

Source: Bureau of Energy (2011).

Table 4. Fraction of Carbon Stored

Fuel	Fraction of Carbon Oxidized
Coal	0.98
Oil and Oil Products	0.99
Gas	0.995
Petroleum Coke	0.99

Source: IPCC (1997a). Revised 1996 IPCC Guidelines for National Greenhouse Gas Inventories: Reference manual Volume 3 Chapter 1.
http://www.ipcc-nggip.iges.or.jp/public/gl/guidelin/ch1ref1.pdf

Table 5. Oxidation Factors as given in the 1996 IPCC Guidelines

5) Convert emissions of carbon to the full molecular weight of CO_2.
Since the ratio of the molecular weight of CO_2 (m.w. 44) to the molecular weight of carbon (m.w. 12) is 44/12, i.e., one ton of carbon is equal to 44/12 tons of CO_2, all final estimates are

multiplied by 44/12 to convert net carbon emissions from energy consumption to total CO_2 emissions.

6) Estimate CO_2 emissions from electricity consumption.
In estimating CO_2 emissions, the Bureau of Energy in Taiwan provides a supplemental method which accounts for all fossil fuel combusted and all electricity consumption. That is, to estimate the CO_2 emissions in a given subsector, the emissions from fossil fuel combustion may be added to the emissions from electricity consumption. Following the approach of the Bureau of Energy, we first sum up electricity consumption from both the energy sector and the non-energy sector, and then distribute the total emissions in kWh across "end-use subsectors," according to the ratio of each sub-sector's electricity consumption to total electricity consumption.

3. Estimation Results

The purpose of this paper is to investigate CO_2 emissions from fossil fuels combustion across Taiwan's 57 subsectors (or industries). The estimates of CO_2 emissions and the related results are presented in the following section.

3.1 Level of Total CO_2 Emissions

Figure 1 shows that between 2005 and 2007, Taiwan's carbon dioxide emissions rose from approximately 250.3 million metric tons to 261.1 million metric tons and the corresponding per capita value rose from 10.99 metric tons to 11.37 metric tons. Thereafter, CO_2 emissions from fossil fuels combustion in 2008 and 2009 show a trend of decrease. Three main reasons contribute to this negative growth. The first may be the economic recession caused by the global financial crisis. The second is that energy consumption went down after the prices of oil and electricity were rationalized. The third is that the government is vigorously conducting related policies and measures on energy saving and carbon reduction. However, the total CO_2 emissions rose again in 2010 from 238.1 million metric tons to 252.9 million metric tons and CO_2 emissions per capita rose from 10.3 metric tons to 10.92 metric tons.

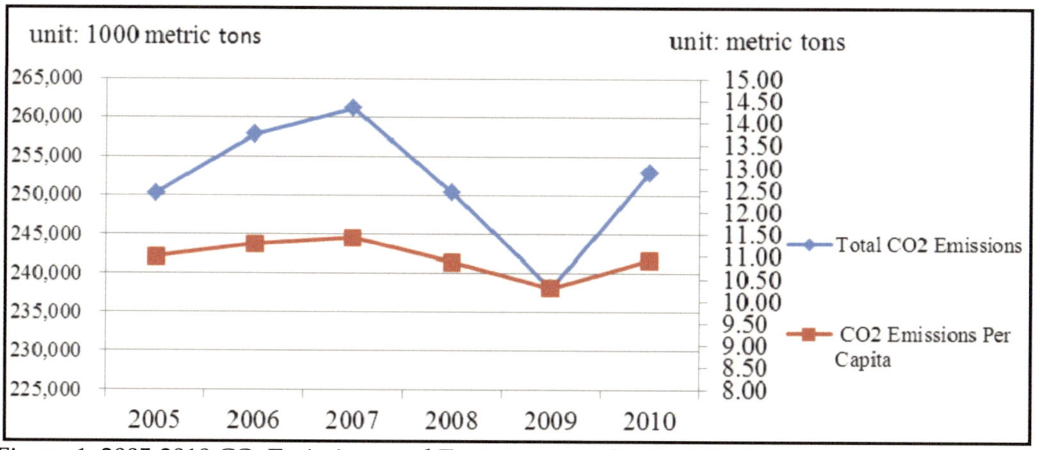

Figure 1. 2005-2010 CO_2 Emissions and Emissions per Capita in Taiwan

3.1.1 CO_2 Emission by Sector

If total CO_2 emissions from fossil fuels combustion are allocated to the economic sector, then the industrial sector emitted 46.4% (116,857,500 metric tons) of subtotal average CO_2 (251,783,300 metric tons) from fuel combustion, the transportation sector 14.1% (35,402,000 metric tons), the service sector14.0% (35,194,170 metric tons), the energy sector11.0% (27,606,330 metric tons), the residential sector12.7% (33,558,590 metric tons), and the agricultural sector 1.3% (3,164,270 metric tons).

Table 6 and Figure 2 show that the industrial sector dominated fossil fuel CO_2 emissions from 2005 to 2010.

Unit: 1000 metric tons

Sector	2005	2006	2007	2008	2009	2010	Average
Energy	28722.5 (11.5%)	29414.79 (11.4%)	29075.4 (11.1%)	26670.48 (10.7%)	25082.73 (10.5%)	26672.05 (10.5%)	27606.33 (11.0%)
Industrial	112628.1 (45.0%)	118306.34 (45.9%)	123527.7 (47.3%)	117509.5 (46.9%)	108637.2 (45.6%)	120536.3 (47.7%)	116857.5 (46.4%)
Transportation	36799.12 (14.7%)	36752.32 (14.3%)	35604.08 (13.6%)	33813.01 (13.5%)	34146.77 (14.3%)	35299.42 (14.0%)	35402.45 (14.1%)
Agricultural	4270.9 (1.7%)	3405.77 (1.3%)	2857.24 (1.1%)	3107.94 (1.2%)	2703.94 (1.1%)	2639.8 (1.0%)	3164.27 (1.3%)
Services	34474.35 (13.8%)	36081.85 (14.0%)	35942.53 (13.8%)	35677.02 (14.2%)	34212.58 (14.4%)	34776.67 (13.7%)	35194.17 (14.0%)
Residential	33447.02 (13.4%)	33851.27 (13.1%)	34133.79 (13.1%)	33592.29 (13.4%)	33305.59 (14.0%)	33021.59 (13.1%)	33558.59 (13.3%)
Total	250342 (100%)	257812.34 (100%)	261140.7 (100%)	250370.3 (100%)	238088.8 (100%)	252945.8 (100%)	251783.3 (100%)

Table 6. Aggregate Fossil Fuel CO2 Emission by Sector, 2005-2010

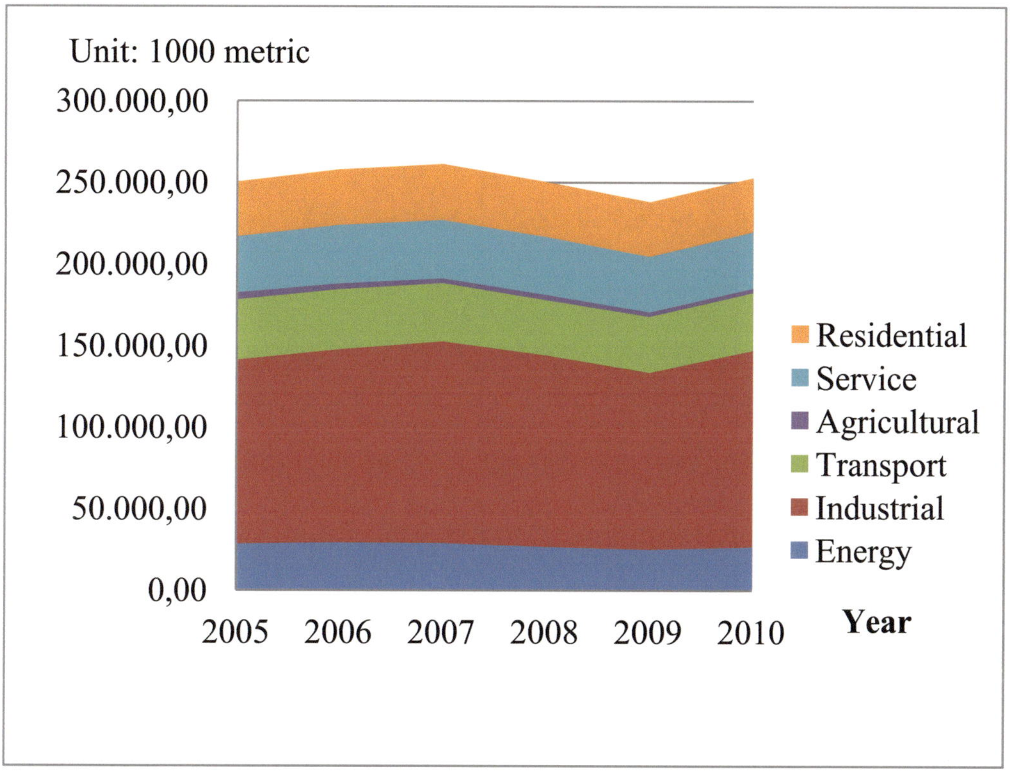

Figure 2. Taiwan's CO2 Emissions from Fuel Combustion, 2005-2010

3.1.2 CO_2 Emissions by Subsector

After further estimating CO_2 emissions from fuel combustion for each subsector, we obtain some important points on the industrial, transport, and energy sectors which need to be addressed.

Industrial sector

Aggregate data from 2005 to 2010 shows that around 46.4% of carbon dioxide emissions are attributable to the industrial sector. Electrical and Electronic Machinery, Petrochemical Materials, Iron and Steel and Artificial Fibers account for around one-half of this amount. CO_2 emissions from the industrial sector are presented in Table 7. The distribution of CO_2 emissions related to the industrial sub-sectors is shown in Figure 3.

Unit: metric tons

Sub-sector	2005	2006	2007	2008	2009	2010	Average
Mining and Quarrying	366,545	366,948	367,046	376,121	381,374	435,224	382,210
Food, Beverage and Tobacco	3,490,823	3,480,565	3,400,271	3,249,925	3,237,041	3,348,114	3,367,790
Textile, Wearing Apparel and Accessories	8,532,973	7,848,310	7,479,042	6,538,729	5,649,643	5,914,709	6,993,901
Leather and Fur	313,297	289,741	275,930	272,655	284,767	257,659	282,341
Wood, Bamboo and Furniture	331,688	338,114	335,333	320,223	283,124	301,477	318,326
Pulp, Paper and Paper Product	4,370,238	4,254,829	4,212,811	3,888,590	3,630,128	3,783,437	4,023,339
Printing	371,902	379,051	385,078	377,819	352,131	364,755	371,789
Basic Industrial Chemicals	3,059,326	3,291,378	3,208,213	2,509,519	2,342,075	2,397,367	2,801,313
Petrochemical Materials	13,716,142	14,952,806	18,630,290	17,503,780	16,994,752	18,591,606	16,731,563
Chemical Fertilizers	1,142,978	1,120,812	1,221,010	1,273,419	1,193,491	1,123,055	1,179,127
Artificial Fibers	7,462,358	7,944,708	8,537,250	7,495,414	7,045,735	7,839,318	7,720,797
Resin, Plastics and Rubber	4,850,328	5,580,850	5,704,723	5,161,846	4,887,555	5,425,261	5,268,427
Other Chemical Materials	1,258,578	1,301,359	1,234,661	922,640	1,034,539	1,153,908	1,150,947
Chemical Products	2,600,729	2,548,404	2,580,231	2,496,558	2,361,979	2,715,784	2,550,614
Rubber Products	934,244	915,705	940,179	930,477	864,426	999,875	930,818

Plastic Products	4,044,185	4,179,991	4,227,114	3,968,155	3,780,430	4,080,562	4,046,739
Cement and Cement Products	7,978,782	7,823,317	7,260,074	6,759,060	5,775,455	5,905,550	6,917,040
Others	3,598,231	3,930,897	3,980,277	3,999,902	3,681,233	4,305,275	3,915,969
Iron and Steel	14,504,718	15,627,904	15,584,547	14,903,393	12,931,939	15,738,728	14,881,871
Non-metal	1,108,608	1,090,043	1,107,395	1,034,513	871,250	988,295	1,033,351
Fabricated Metal Products	4,231,155	4,388,725	4,478,430	4,324,757	3,704,740	4,452,390	4,263,366
Machinery and Equipment	1,148,366	1,217,632	1,255,934	1,243,187	1,049,711	1,292,931	1,201,294
Electrical and Electronic Machinery	18,404,777	20,356,701	21,846,391	22,822,217	21,596,448	24,100,092	21,521,104
Transport Equipment	1,842,836	1,893,038	1,976,473	1,941,264	1,847,092	2,112,642	1,935,557
Precision Instruments	402,450	604,739	723,056	722,500	567,607	584,551	600,817
Miscellaneous Industries	1,048,149	1,058,022	1,043,500	963,693	867,631	896,664	979,610
Water Supply	846,637	897,638	917,484	904,818	862,141	869,290	883,002
Construction	667,115	624,138	614,975	604,369	558,787	557,767	604,525
Total	112,628,153	118,306,364	123,527,720	117,509,541	108,637,223	120,536,284	116,857,547

Table 7. CO_2 Emissions from Industrial Sector, 2005-2010

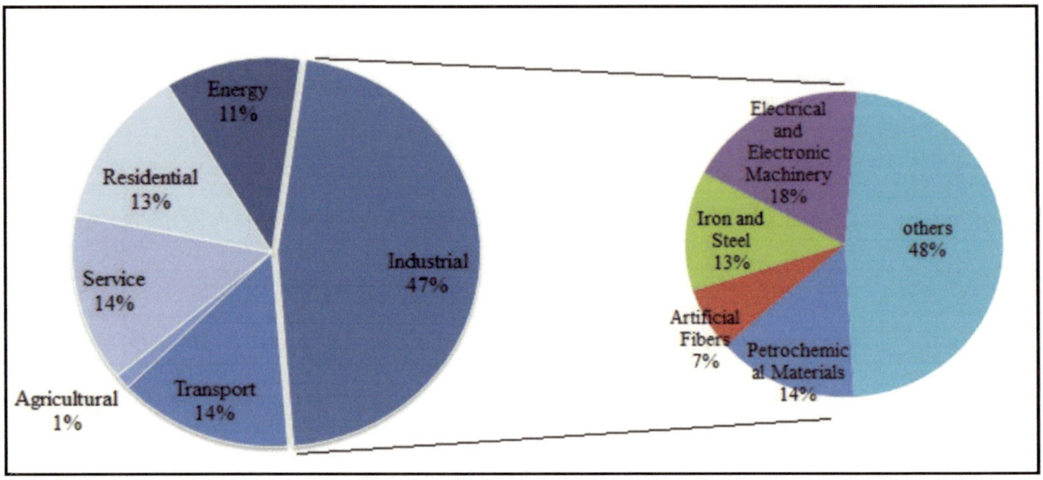

Figure 3. CO_2 Emissions from Industrial Sector, 2005-2010

Transport Sector

The transport sector, the second largest emitting sector, contributes 14.1% of the total CO_2 emissions in Taiwan. CO_2 emissions from the transport sector are shown in Table 8. In the transport sector, road transport is responsible for a significant share of the CO_2 emissions as shown in Figure 4.

Unit: metric tons

Sub-sectors	2005	2006	2007	2008	2009	2010	Average
Domestic Air	586,101.1	510,875.6	386,334.7	256,321.4	225,937.9	228,419.8	365,665.1
Road	34,644,467.5	34,714,511.4	33,612,956.1	31,955,343.2	32,330,742.1	33,375,030.1	33,438,841.7
Rail	446,111.9	469,986.7	654,017.4	831,505.4	807,533.1	839,938.7	674,848.9
Internal Navigation	1,122,435.9	1,056,947.8	950,773.8	769,841.7	782,556.6	856,034.7	923,098.4
Total	36,799,116.4	36,752,321.5	35,604,082.0	33,813,011.7	34,146,769.7	35,299,423.3	35,402,454.1

Table 8. CO_2 Emissions from the Transport Sector, 2005-2010

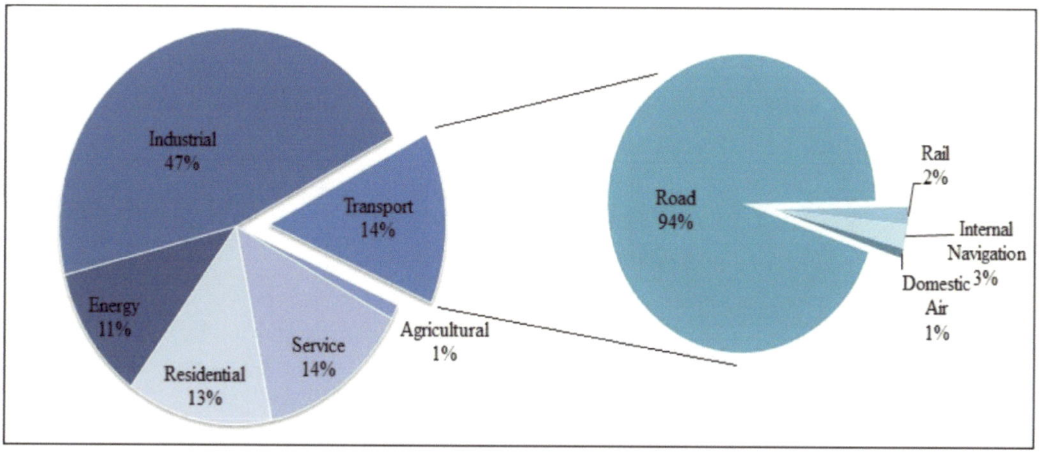

Figure 4. CO_2 Emissions from Transportation Sector, 2005-2010

The fact that road activity generates the most CO_2 emissions means that road vehicles, including motorcycles, passenger cars and trucks, account for approximately 94% of all transport-related CO_2 emissions. This percentage is much higher than that of EU 27, which can be seen from Figure 5.

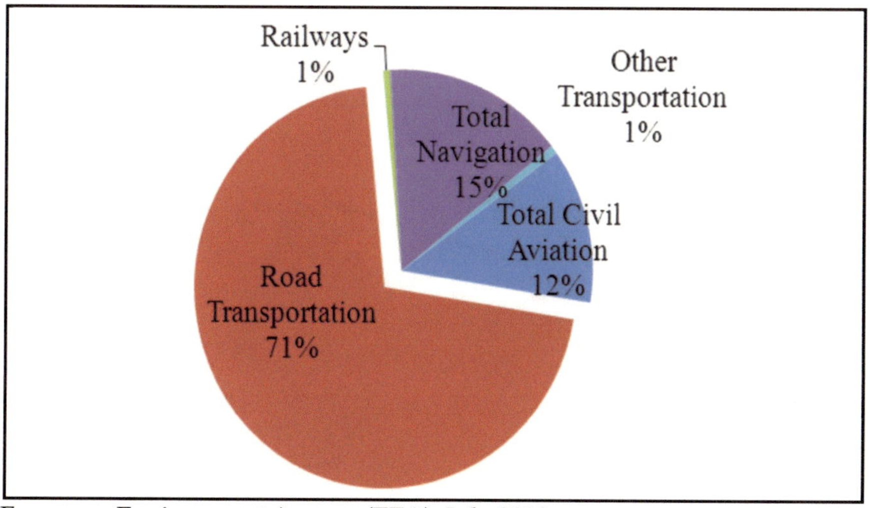

Source: European Environment Agency (EEA), July 2009

Figure 5. Share by Mode in Total Transport CO_2 Emissions, including International Bunkers : EU27(2007)

Energy Sector

The energy sector, the fifth largest CO_2 emitter, accounts for 11.0% of total CO_2 emissions. From 2005 to 2010, the petroleum refineries industry accounted for the major share (35% on average) of total CO_2 emissions in the energy sector, followed by electricity plants and blast furnaces. CO_2 emissions from the energy sector are presented in Table 9.The distribution of CO_2 emissions related to the energy sector is shown as Figure 6.

Unit: metric tons

Sub-Sectors	2005	2006	2007	2008	2009	2010	Average
Coal Mines	5,897.5	2,583.1	5,688.6	5,038.5	6,022.2	5,082.7	5,052.1
Coke Ovens	3,079,585.4	3,158,183.8	3,134,733.0	2,951,828.4	2,530,175.0	3,388,922.5	3,040,571.4
Blast Furnaces	3,572,462.4	3,866,130.9	4,043,693.0	3,920,091.1	3,385,142.1	4,227,864.3	3,835,897.3
Oil and Gas Extraction	58,481.1	57,234.1	45,683.8	10,857.1	10,705.7	9,505.5	32,077.9
Petroleum Refineries	11,187,078.6	10,776,132.1	10,151,043.1	8,792,185.8	8,690,228.1	8,783,988.3	9,730,109.3
Electricity Plants	6,280,975.6	6,614,605.4	6,740,893.6	6,599,869.4	6,200,770.3	6,320,543.7	6,459,609.7
Pump-Generated Electricity	3,090,802.0	3,201,415.9	3,020,076.4	2,656,614.9	2,663,217.0	2,346,080.3	2,829,701.1
Cogeneration Plants	1,028,417.7	1,257,728.3	1,437,248.7	1,183,130.8	1,313,200.3	1,344,551.6	1,260,712.9
Gas Companies	418,796.7	480,773.4	496,344.1	550,863.7	283,266.0	245,509.4	412,592.2
Total	28,722,497.0	29,414,787.0	29,075,404.3	26,670,479.7	25,082,726.7	26,672,048.3	27,606,323.9

Table 9. CO2 Emissions from the Energy Sector, 2005-2010

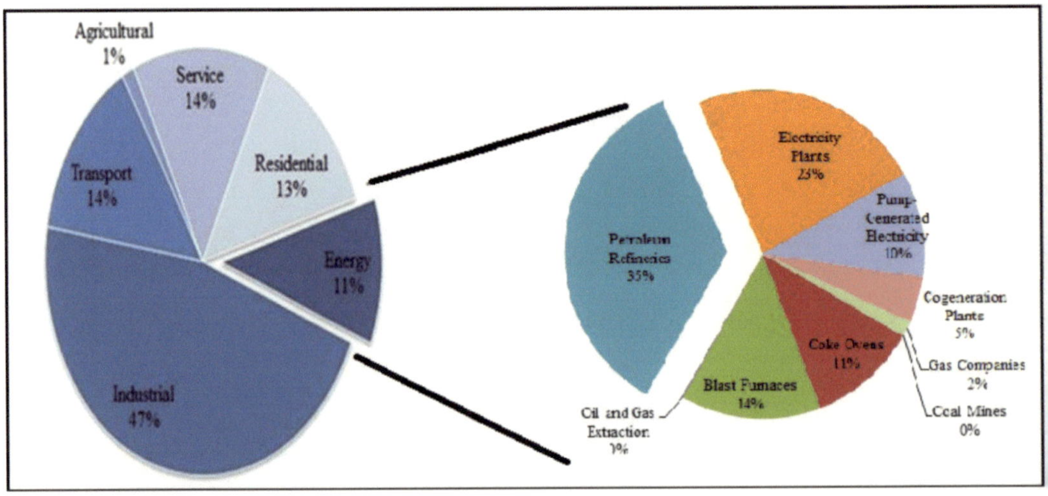

Figure 6. CO_2 Emissions from Energy Sector, 2005-2010

3.2 10 High-Emitting Subsectors in Taiwan

The residential sector in Taiwan is the largest contributor to CO_2 emissions. As discussed earlier, road transport is responsible for a significant share of the CO_2 emissions in the transport sector. It also is ranked second in fossil fuel CO_2 emissions among 57 sub-sectors. Electrical and electronic machinery industry is ranked third.

The top 10 high-emitting subsectors in Taiwan are presented in Table 10 and the time trend of the 10 high-emitting subsectors in Taiwan is shown in Figure 7.

Unit: metric tons

Ranking	Average CO2 Emissions
Residential	33,558,590.7
Road	33,438,841.7
Electrical and Electronic Machinery	21,521,104.3
Petrochemical Materials	16,731,562.8
Iron and Steel	14,881,871.4
Other Services	12,219,482.9
Petroleum Refineries	9,730,109.3
Artificial Fibers	7,720,797.0
Textile, Wearing Apparel and Accessories	6,993,901.0
Cement and Cement Products	6,917,039.71

Table 10. Top 10 High CO_2 Emitters in Taiwan, 2005-2010

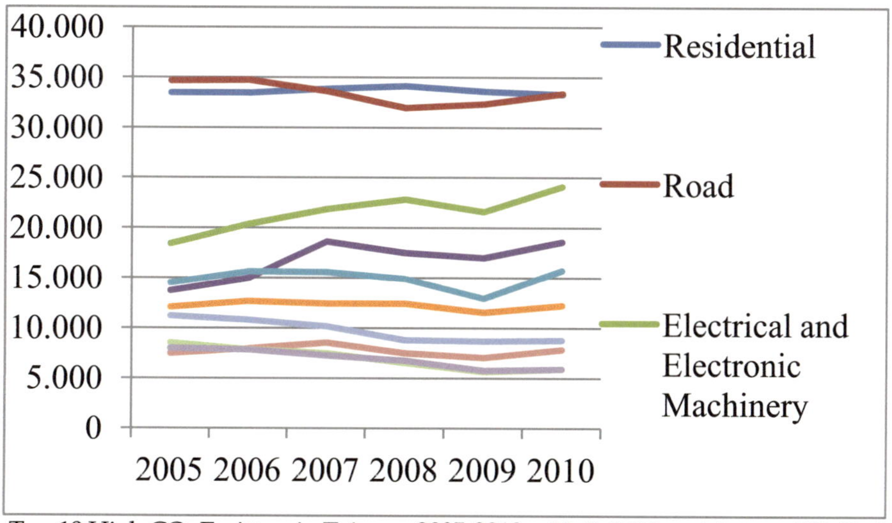

Figure 7. Top 10 High CO_2 Emitters in Taiwan, 2005-2010 Unit: 1000 metric tons CO_2 Emissions

4. Conclusion and Future Research Direction

This paper is part of an ongoing research project designed to investigate the potential size of Taiwan's carbon market. When tackling this big issue on the size of the carbon market, we first use IPCC's sectoral approach to estimate CO_2 emissions from fuel combustion and examine the sectoral and subsectoral distribution of CO_2 emissions in Taiwan. Utilizing the Energy Balance Sheet compiled by the Bureau of Energy, this analysis is based on the fuel consumed in each subsector and the electricity used in the subsector.

With the results obtained in this paper, we are planning to examine the demand and supply structure of Taiwan's carbon market by projecting CO_2 emission data to year 2012 and 2013. Since the cap (emission rights) is given, the quantity of demand for emission rights and the quantity of supply for emission rights could thus be identified.

5. References

[1] Bureau of Energy, Ministry of Economic Affairs (2010), *Analysis of Fossil Fuel-related CO_2 Emissions in Taiwan*, http://www.moeaboe.gov.tw/promote/greenhouse/PrGHMain.aspx?PageId=pr_gh_list.

[2] Bureau of Energy, Ministry of Economic Affairs, *Energy Balance Sheet*, http://www.moeaboe.gov.tw/English/Statistics/EnStatistics.aspx.

[3] European Environment Agency (2009), *EU Energy in Figures 2010 CO_2 Emissions from Transport by Mode.* http://ec.europa.eu/energy/publications/doc/statistics/ext_co2_emissions_from_transport_by_mode.pdf.

[4] IPCC (1997a). Revised 1996 IPCC Guidelines for National Greenhouse Gas Inventories: Reference manual Volume 3 Chapter 1. http://www.ipcc-nggip.iges.or.jp/public/gl/guidelin/ch1ref1.pdf.

[5] IPCC (1997b). Revised 1966 IPCC Guidelines for National Greenhouse Gas Inventories: Workbook (Volume 2), http://www.ipcc-nggip.iges.or.jp/public/gl/guidelin/ch1wb1.pdf.

5

A Preliminary Look at the Relationship Between Environmental Change and Economic Growth in Taiwan

Kuang-Ta Lo and Ya-Ting Yang
Department of Public Finance, National Chengchi University
Taiwan

1. Introduction

The relationship between environmental change and economic growth is two-sided. On the one hand, economic development leads to a change of environmental quality. This phenomenon is usually described by the well-known environment Kuznets curve (EKC). The EKC illustrates that economic growth and environmental degradation have an inverted U-shaped relationship. During the initial industrial period, people usually care more about their jobs and a better income than they do about the issue of environment; therefore, pollution tends to grow rapidly in this period. When income increases to a certain level, people begin to pay more attention to the quality of the environment so that at higher-income levels economic growth leads to environmental improvement. In other words, the EKC hypothesis contends that pollution increases initially as a country develops its industry and thereafter declines after reaching a certain level of economic progress (Stern, 1996).

On the other hand, a changing environment also affects economic growth. Environmental change can be observed from changes of temperature, in levels of precipitation, CO2 and SO2, etc. The Intergovernmental Panel on Climate Change (IPCC) Report argues that the effects from environmental change will impact all countries of the Southeast Asia region.[8] According to IPCC, crop yields could decrease up to 30% in central and south Asia by the middle of this century. The rapid population growth and urbanization in the region will magnify the number of people malnourished and subject to the risk of hunger due to climate change. More specifically, a recent study estimates a 2~5% decrease in yield potential of wheat and maize for a temperature rise of 0.5 to 1.5°C in India.[9] Dell et al. (2009) point out that from their research of cross-sectional data, a national income per-capita will fall 8.5% on average per degree Celsius rise in temperature, suggesting a simple method to calculate

---

[8] http://www.searo.who.int/en/Section260/Section2468/Section2500_14162.htm
[9] Aggarwal, P. (2003). Impact of climate change on Indian agriculture. J. Plant Biol., 30: 189-198.

how warming might influence future standards of living. The Asian Development Bank (ADB) also states that climate change may have a major impact on economic growth in Asia. The results of a recent ADB study about climate change in Southeast Asia show the total cost of losses due to climate change is quite large. If nothing is done, then the total cost of climate change for the countries of Indonesia, the Philippines, Thailand, and Vietnam could reach a combined 6.7% of GDP every year until 2100.[10]

Understanding the impact of climate change on the world economy is, obviously, of paramount importance for both climate change mitigation and adaptation policies. However, as Rosen and Mensbrugghe (2010) argue, modeling climate impact is a challenging undertaking for two main reasons. First, climate change is a systemic phenomenon in terms of both natural and human systems. In the so-called "Earth System," physical elements like the oceans, winds, the stratosphere, etc., interact in the determination of global climate conditions. In terms of socio-economic consequences, market linkages and trade propagate the effects of noneconomic factors throughout the globalized economy. As climate change is an intrinsically systemic phenomenon, it is inherently affected by complexity and uncertainty. Second, socio-economic impacts of climate change have different dimensions (e.g., sea level rise, human health, et al.), each one with different mechanisms and implications. To achieve a realistic assessment of the impacts, it is necessary to separately and adequately address each dimension.

Weather patterns exhibit large variations in amplitude and intensity. When unusual volumes of rainfall or extremes in air temperatures occur in populated regions, humans often suffer serious economic crises and loss of life. There is evidence suggesting that human activities are influencing climate on a large scale. One major effect has been the enhanced greenhouse effect related to massive increases in the amounts of airborne CO2, CH4 and N2O, between 1750 and 2005 (IPCC). As observed on Mauna Loa, Hawaii, the amount of CO2 has increased since 1958 at a rate of 1.5 parts per million (ppm) per year. Meanwhile, the global mean surface temperature increased 0.74°C±0.18°C between 1906 and 2005 (IPCC). In 2005, the global annual mean surface temperature was the warmest among temperature observations made since 1880 (Shein et al., 2006).

Taiwan is a densely populated island. The development of its industry has necessitated ever-increasing quantities of fossil fuel. On the other hand, the emission of greenhouse gases also has been growing quickly. These gases will continue to accumulate as time goes by and will indirectly influence climate change. Weather thus could be influenced by both natural and human factors. Moreover, bizarre weather phenomena such as maximum temperatures, minimum temperatures and diurnal temperature changes (DTC) often cause great damage and loss of crops, fishery and so on. Thus, the primary purpose of this research is to understand how the environment has been changed in Taiwan over the past years. This study further examines the correlation between environmental changes and economic growth in Taiwan. Additionally, we also analyze whether there are any spatial variations of environmental changes in different areas.

---

[10] http://www.banningbusinesscenter.com/climate-change-threatens-asian-economic-growth.html.

2. Issues in the Literature

The causal chain linking economic behavior today to economic consequences tomorrow via environmental changes can be summarized as follows:

Economic activities → Emissions → Atmospheric Concentrations → Radiative Forcing → environment or climate changes → impacts on physical and ecological systems → impacts on or damage to economies.

Each of the linkages between the components above involves complex factors.[11] For example, the link between economic activities and emissions involves population change, rates of economic growth, the specific stage of economic development (e.g., an economy reliant on heavy industry versus a service-based economy), the type of energy used to fuel the economy, energy efficiency (the amount of energy used to produce a unit of GNP changes as an economy develops), and technology. In addition, it is affected by the way global incomes accrue across countries. Hence, emissions do not have any simple proportional relationship to economic activity. As far as links from emissions to atmospheric concentrations of greenhouse gases and from concentrations to temperature change are concerned, what matters is the amount of greenhouse gases in the atmosphere. Annual emissions do not therefore have any simple proportional link to concentrations. Annual emissions add to the overall amount and the stored emissions (the atmospheric concentration) also decay at various rates. Most importantly, it is the overall amount that helps to determine temperature change. Even here the link is complex because the change in "radiative forcing" is not proportional to concentrations. The link from temperature change to economic damage depends on a further set of factors: how economies adapt to temperature change, how vulnerable economies are, how rapid warming is, and whether there are abrupt changes in temperature and weather events.

Environmental change manifests itself in increases of temperature, fluctuation of precipitation, rises in sea level, and the intensification of natural hazards, such as storms, floods, droughts, and landslides. One major consequence of global warming could be greater scarcity and variability of renewable resources in many parts of the world. With increasing concerns about such global effects of climate change a group of scholars, commonly referred to as neo-Malthusians, has posited that climate change is a threat to international security because it could increase resource scarcity (Schubert et al., 2009; Homer-Dixon, 1999; Homer-Dixon & Blitt, 1998 Bächler et al., 1996). However, other scholars, commonly referred to as cornucopians or resource optimists, do not share this pessimistic view. They believe that humanity can adapt to increasing resource scarcity through appropriate market mechanisms (pricing), technological innovation, and other means (Lomborg, 2001; Simon, 1998).

The economies of some countries are more vulnerable to climate change than the global average. Developing countries in general have a larger share of their economies in

[11] House of Lords (2005), The Economics of Climate Change.

agriculture and forestry. They also tend to be in the low latitudes where the impacts to these sectors will be the most severe. The low latitudes tend to be too hot for the most profitable agricultural activities and any further warming will further reduce productivity. Up to 80% of the damages from climate change may be concentrated in low-latitude countries (Mendelsohn et al. 2006).

Mendelsohn (2009) has argued that some damages from environment change will not affect the global economy, but will simply reduce the quality of life. Ecosystem change will result in massive shifts around the planet. Some of these shifts are already being reflected in agriculture and timber but they go beyond the impacts to these market sectors. Parks and other conservation areas will change. Animals will change their territorial range. Endangered species may be lost. Although these impacts will likely lead to losses of nonmarket goods, it is hard to know what value to assign to these effects. Another important set of nonmarket impacts involves health effects. Heat stress may increase. Vector-borne diseases may extend beyond current ranges. Extreme events could threaten lives. All of these changes could potentially affect many people if we do not adapt. However, it is likely that public health interventions could minimize many of these risks. Many vector-borne diseases are already controlled at relatively low cost in developed countries. Heat stress can be reduced with a modicum of preventive measures. Deaths from extreme events can be reduced by a mixture of prevention and relief programs. As the world develops, it is likely that these risks may involve higher prevention costs, but not necessarily large losses of life.

Furthermore, Dell et al. (2008) argue that higher temperatures have large, negative effects on economic growth, but only in poor countries. In poor countries they estimate that a 1°C temperature increase in a given year reduced economic growth in that year by about 1.1%. In rich countries, changes in temperature had no discernable effect on growth. Changes in precipitation had no substantial effects on aggregate output in either poor or rich countries. Since they find no effects on rich countries, their results thus further imply that future climate change may substantially widen income gaps between rich and poor countries.

3. Relevant Background of Taiwan

3.1 Geographic Features and Natural Resources

Taiwan occupies an area of 36,191 square kilometers, somewhat smaller than the size of the Netherlands (41,526 km2). At mid-2010, its population stood at 23.1 million. With 638 persons per square kilometer, Taiwan is one of the most densely populated areas in the world. Three-quarters of the land is mountainous, with a spine-like ridge of steep mountains extending from north to south. About 60% of the land is forested, but forest resources are minimally exploited because of limited accessibility and environmental concerns. Even though only one-quarter of the land is arable, the subtropical climate permits multi-cropping of rice and growing of fruit and vegetables all year round. However, agricultural production accounted for only 1.7% of gross domestic product (GDP) in 2009. Although Taiwan does have deposits of coal, limestone, marble, dolomite, and natural gas, it is not richly endowed by nature. Indeed, more than 90% of its energy needs are met by imports, and its rapid industrialization also has relied heavily on imports of raw materials.

However, Taiwan has an ample supply of human resources, of which it has made highly effective use.[12]

3.2 Economic Background

Through decades of hard work and sound economic management, Taiwan has transformed itself from an underdeveloped, agricultural island to an economic power that is a leading producer of high-technology goods. The first stage of Taiwan's economic development extended from 1952 through 1980. During this period, Taiwan averaged an annual economic growth rate of 9.21%, which was the highest in the world. In 1962, agriculture lost its key position as the driving force behind Taiwan's economy, making way for the rapidly developing industrial sector. With the exception of two energy crises, in 1973-1974 and 1979-1980, Taiwan's industries maintained an average annual growth rate of around 14%.[13]

The second stage of Taiwan's economic development ran from 1981 through 2000. During this period, economic conditions around the world and within Taiwan itself underwent great changes. Combined external and internal forces exerted a rather detrimental effect on Taiwan's economic development, slowing the growth rate to a low of 7.15%. The focus of Taiwan's economy slowly shifted from the industrial sector to the service sector. Meanwhile, the agricultural sector grew a mere 0.63% annually as its GDP share continued to diminish. Limited natural resources and a high population density mean that Taiwan is not self-sufficient. Hence, foreign trade has come to play a leading role in Taiwan's economic development. The development of foreign trade and the increase of foreign investment are driving forces behind Taiwan's industrial sector, which in turn fuels development in the service sector. From 1952 through 1980, the annual growth rate of commodity and labor exports averaged 16.5%, while local demand grew an average of 10.97% per year. During the second stage of economic development, commodity and labor exports grew 10.05% per year, while local demand grew 7.51%. From these figures, the importance of foreign trade to Taiwan's economic development can be seen quite clearly.

Although Taiwan enjoyed sustained economic growth, full employment, and low inflation for many years, in 2001, Taiwan joined other regional economies in its first recession since 1949. From 2002 to 2007, Taiwan's economic growth ranged from 3.5% to 6.2% per year. With the global economic downturn, Taiwan's economy slumped into recession in the second half of 2008. Its real GDP, following growth of 5.98% in 2007, rose 0.73% in 2008 and contracted 1.93% in 2009. The economy began to recover in 2010 and the GDP grew remarkably by 10.88% in 2010.[14] Some economic performance indicators of Taiwan from 1978 to 2010 are provided in Table 1.

---

[12] http://www.cepd.gov.tw/m1.aspx?sNo=0014790&ex=2&ic=0000153
[13] http://www.gio.gov.tw/info/taiwan-story/economy/edown/3-5.htm
[14] http://www.traveldocs.com/tw/economy.htm

	Economic Growth Rate	Per capital GDP	Unemployment Rate	Value of Export	Exchange Rate
	%	USD	%	USD	NTD/1USD
1978	13.49	1,599	1.67	12,755	36.94
1979	8.01	1,943	1.27	16,169	36.00
1980	7.32	2,385	1.23	19,878	36.78
1981	6.46	2,730	1.36	22,686	37.79
1982	3.97	2,703	2.14	22,297	39.86
1983	8.32	2,902	2.71	25,207	40.22
1984	9.32	3,219	2.45	30,580	39.42
1985	4.07	3,290	2.91	30,819	39.8
1986	11	4,007	2.66	39,931	35.45
1987	10.68	5,265	1.97	53,754	28.5
1988	5.57	6,146	1.69	60,784	28.12
1989	10.28	7,558	1.57	66,435	26.17
1990	6.87	8,124	1.67	67,425	26.88
1991	7.88	9,016	1.51	76,563	25.7
1992	7.56	10,625	1.51	82,122	25.37
1993	6.73	11,079	1.45	85,957	26.62
1994	7.59	11,982	1.56	94,300	26.16
1995	6.38	12,918	1.79	113,342	27.22
1996	5.54	13,428	2.6	117,581	27.44
1997	5.48	13,810	2.72	124,170	32.52
1998	3.47	12,598	2.69	112,595	32.16
1999	5.97	13,585	2.92	123,733	31.34
2000	5.8	14,704	2.99	151,950	32.96
2001	-1.65	13,147	4.57	126,314	34.94
2002	5.26	13,404	5.17	135,317	34.71
2003	3.67	13,773	4.99	150,601	33.92
2004	6.19	15,012	4.44	182,370	31.68
2005	4.7	16,051	4.13	198,432	32.78

	Economic Growth Rate	Per capital GDP	Unemployment Rate	Value of Export	Exchange Rate
	%	USD	%	USD	NTD/1USD
2006	5.44	16,491	3.91	224,017	32.55
2007	5.98	17,154	3.91	246,677	32.39
2008	0.73	17,399	4.14	255,629	32.81
2009	-1.93	16,353	5.85	203,675	31.98
2010	10.88	18,588	5.21	274,601	30.32

Table 1. Economic Outlook of Taiwan, 1978-2010

However, behind this image of economic achievement, the level of economic development varies significantly from one region to another (Hou, 2000). Taiwan is typically divided into five geographic units—North, Central, South, East and the outlying islands (Figure 1). The northern region with Taipei at its center is the most urbanized and populated. The central and southern regions, punctuated by a few major cities, have been predominantly agricultural, but are now rapidly becoming industrialized. The eastern region, known for its rugged coastal landscape and poor accessibility, remains largely excluded from major development. The outlying islands have also been excluded from economic development. In terms of population, the northern region accounted for 44.63% of the national population of 23.05 million in 2010. In particular, Taipei City and Taipei County together have more than a quarter of the entire population in Taiwan (28.26%). The southern region and the central region have 27.91% and 24.99%, respectively. The eastern region only has 2.47%.

Regional socioeconomic differences are clearly noticeable through comparisons of household income between counties and cities. With the exception of highly urbanized Taipei, Hsinchu, and Taoyuan Counties where many of Taiwan's high-tech firms are located, the average disposable household income is consistently lower than that of the major urban areas. The average family income in the poorest area, Yunlin County, is only 54.24% of the average income in Hsinchu County, and only approximately half of that of Taipei City in 2009 (49.35%).

Concentration of firms corresponds with the regional population breakdown. Almost 46.7% of firms are located in the northern region, while 26.1% and 24.3% of firms are located in the southern and central regions, respectively. The eastern region and the outlying islands account for less than 3%. Besides, patterns of GDP per capita and concentration of firms largely correspond with the urbanization and industrialization of the area. The counties with higher percentages of employment in agriculture, fisheries, and mining, and with a larger area of cultivated land tend to have lower GDP per capita. Administrative resources at the county level also largely correspond to the wealth of the regions. Table 2 provides some economic outlook of four major cities in Taiwan.

3.3 Environmental Change

In the past 100 years, Taiwan experienced an island-wide warming trend (1.0-1.4°C/100 years). The warming in Taiwan is closely connected to a large-scale circulation and surface air temperature (SAT) fluctuations, such as the "cool ocean warm land" phenomenon. The water vapor pressure has increased significantly and may have resulted in a larger temperature increase in summer. The probability for the occurrence of high temperatures has increased and the result suggests that both the mean and variance in the SAT in Taiwan have changed significantly since the beginning of the 20th century. Although, as a whole, the precipitation in Taiwan has shown a tendency to increase in northern Taiwan and to decrease in southern Taiwan in the past 100 years, it exhibits a more complicated spatial pattern. The changes occur mainly in either the dry or rainy season and result in an enhanced seasonal cycle. The changes in temperature and precipitation are consistent with the weakening of the East Asian monsoon (Hsu and Chen, 2002).

More specifically, the annual mean temperature in Taiwan increased significantly during the past century and especially in the past 50 years; the trend of the annual mean minimum temperature was statistically more significant than that of the annual mean maximum temperature (Lai and Cheng, 2010). Only a few studies have been conducted regarding trends in the annual mean maximum temperature, the annual mean maximum temperature, and their differences in Taiwan; furthermore, differences in air temperature changes between urban and rural areas have not been closely discussed. These two subjects are both interesting and important. The daily mean air temperature cannot exactly reflect temperature changes throughout each day, because it is an average of the daily maximum and minimum temperatures. Therefore, the average temperature may underestimate the number and duration of high-heat events.

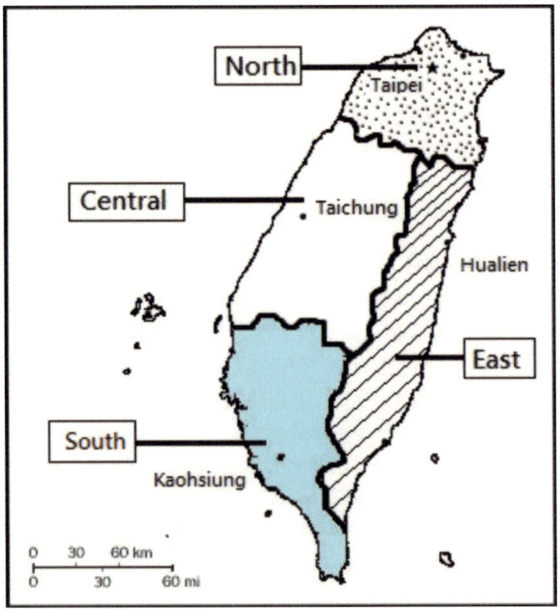

Figure 1. Major geographical areas in Taiwan

	Taiwan			Taipei (North)			Taichung (Central)			Kaohsiung (South)			Hualien (East)		
	DI	UR	UA	DI	UR	UA	DI	UR	UA	DI	UR	UA	DI	UR	UA
1998	231.6	2.6	12.3	331.3	2.6	100.0	231.0	2.6	22.2	226.1	3.1	13.9	213.3	2.8	1.8
1999	244.9	2.9	12.3	336.7	2.9	100.0	250.9	3.1	22.2	235.7	3.7	13.9	209.1	3.7	1.8
2000	246.3	3	12.3	338.2	2.7	100.0	229.3	3.4	22.2	240.1	3.9	13.9	226.2	3.9	1.8
2001	242.6	4.6	12.4	339.3	3.9	100.0	234.2	4.9	22.2	239.3	5	13.9	227.0	5.1	2.7
2002	240.0	5.2	12.5	357.2	4.6	100.0	223.2	5.4	22.5	228.3	5.5	14.0	201.0	5.5	2.7
2003	249.8	5	12.5	365.7	4.6	100.0	237.9	5.3	22.3	237.5	5.3	14.0	220.8	5.3	2.7
2004	254.6	4.4	12.5	380.5	4.2	100.0	224.2	4.6	22.3	249.8	4.6	14.0	236.7	4.8	2.7
2005	261.6	4.1	12.5	392.4	3.9	100.0	232.4	4.2	22.3	260.0	4.2	14.1	228.5	4.4	2.7
2006	267.8	3.9	12.5	378.0	3.7	100.0	246.3	4.1	22.6	259.6	4.1	14.1	256.6	4.2	2.7
2007	273.3	3.9	12.5	389.1	3.7	100.0	261.4	4	22.6	279.5	4.1	14.1	247.3	4.1	2.7
2008	272.7	4.1	12.6	386.3	4	100.0	244.9	4.2	22.6	268.9	4.3	14.1	227.1	4.2	2.7
2009	265.8	5.9	12.7	387.1	5.8	100.0	238.2	5.9	22.6	265.9	5.9	14.2	223.0	5.9	2.7
Average 1998-2009	254.3	4.1	12.5	365.2	3.9	100.0	237.8	4.3	22.4	249.2	4.5	14.0	226.4	4.5	2.5

Table 2. Economic Outlook of Taiwan and Four Major Cities in Taiwan, 1998-2009

Notes: 1. DI is the disposable income per capita, measured in NTD$1,000.

2. UR is the unemployment rate (%).

3. UA is the percentage of urbanized areas (%).

According to the analysis in the Statistics of Climate Changes in Taiwan, recently published by the Central Weather Bureau (CWB), the average temperature in Taiwan of the last 100 years has increased 0.8°C, with an increase of 1.2°C in flat area, 1.4°C in urban area, 0.9°C over western suburban area, 1.3°C in eastern suburban area, 0.6°C for mountain area, and 1.1°C at outlying islands. In urban areas, the increased minimum temperature (2.1°C) is three times that of the area's maximum temperature (0.7°C), which indicates that the increase of night-time temperature is greater than in day time. In terms of the season, the amplitude of temperature increase is larger in spring and autumn.

There are slight changes in the trend of precipitation in the last 100 years in Taiwan. Northern flat area appears to have more rainfall, especially in autumn; while southern Taiwan and mountain area get less rainfall, especially in winter. A decrease in raining hours indicates that the precipitation intensity (precipitation per unit time) has increased. Except in mountain area, the number of days with precipitation over 30 mm has increased over the last 100 years.

4. Empirical Design

4.1 Interest of Data

The purpose of this study is to understand how and to what extent environment changes affect economic growth in Taiwan. The indicators of environment changes used in this study include temperature, precipitation and ozone level (O3).

The existing literature provides significant amounts of evidence that the change of temperature and rainfall will affect economic output.[15] Such evidence also suggests that climate change should affect economic growth. If climate change affected only the level of economic output, for example, by reducing agricultural yields when temperature rises (precipitation falls), this would imply that subsequent temperature decreases (precipitation increases) - due, for example, to stringent abatement of emissions - should return the GDP to its previous level. But this is not the case if climate change affects economic growth. Koubi et al. (2010) provide the following two reasons. First, economic growth will be lower even if GDP returns to its previous level because of forgone consumption and investment due to lower income during the period of higher temperature (lower precipitation). In addition, as long as countries spend some resources to adapt to climate change, they incur opportunity costs in terms of not spending these resources on R&D and capital investment. This has negative effects on economic growth. Moreover, given the short time-series data used in existing research on climate effects on economic conditions, even slightly persistent effects on the level of output will impact on the sample mean of growth. That is, using economic growth rates will also capture the effects on GDP levels. But using the level of GDP instead of its growth rate may miss the effects on the growth rate. For these reasons we concentrate on climate change effects on economic growth.

---

[15] For instance, Mendelsohn et al., 1998; Mendelsohn, Dinar & Williams, 2006; Nordhaus & Boyer, 2000; Tol, 2002; Deschenes & Greenstone, 2007.

In addition to the change of temperature and precipitation, we also consider ozone concentrations. Although the ozone layer in the upper atmosphere is beneficial, preventing potentially damaging electromagnetic radiation from reaching the earth's surface, ozone in the lower atmosphere is an air pollutant with harmful effects on the respiratory systems of animals and will burn sensitive plants. The increase in ozone is of further concern because ozone present in the upper troposphere acts as a greenhouse gas, absorbing some of the infrared energy emitted by the earth. Quantifying the greenhouse gas potency of ozone is difficult because it is not present in uniform concentrations across the globe. However, the most widely accepted scientific assessments relating to climate change (e.g., the Intergovernmental Panel on Climate Change Third Assessment Report) suggest that the radiative forcing of tropospheric ozone is about 25% that of carbon dioxide. However, tropospheric ozone is a short-lived greenhouse gas, which decays in the atmosphere much more quickly than carbon dioxide. Because of its short-lived nature, tropospheric ozone does not have strong global effects, but has very strong radiative forcing effects on regional scales.

The data in this study are mainly collected from the 30 observation stations of the Central Weather Bureau (CWB), and the Directorate-General of Budget, Accounting and Statistics.

4.2 Trend and Correlation Analyses

In this subsection, we will investigate the relationship between environmental changes and economic development in Taiwan and its four major cities. Taipei City has been considered the cultural, economic and political center of Taiwan; it is also the largest city in northern Taiwan. Taichung City, a city of mixed commercial and industrial activities, is the largest city in central Taiwan. Kaohsiung, the largest city in southern Taiwan, has been the heavy industry center of Taiwan. Hualien, located in eastern Taiwan, is the smallest of the four cities, and agriculture is its major economic activity.

From Figure 2.1, it is obvious that the mean temperature in Taiwan increased during the period of 1952-2010. For example, it was 22.75°C in 1955, 23.14 °C in 1975, 23.25 °C in 1995, and 23.73°C in 2005. However, the standard deviation of 12 months per year is smaller (Figure 2.2). It can be further explained by the trends of the maximum and minimum temperatures (Figure 2.3). The figures also indicate that the maximum mean monthly temperature (Tmax) and the minimum mean monthly temperature (Tmin) increased in Taiwan, and that the rate of increase for Tmin was higher than that for Tmax. Thus, TDiff decreased significantly (Figure 2.4). In most areas in Taiwan, Tmin increased more than Tmax, inducing an obvious reduction in TDiff and a smaller variation (standard deviation) over the past years.

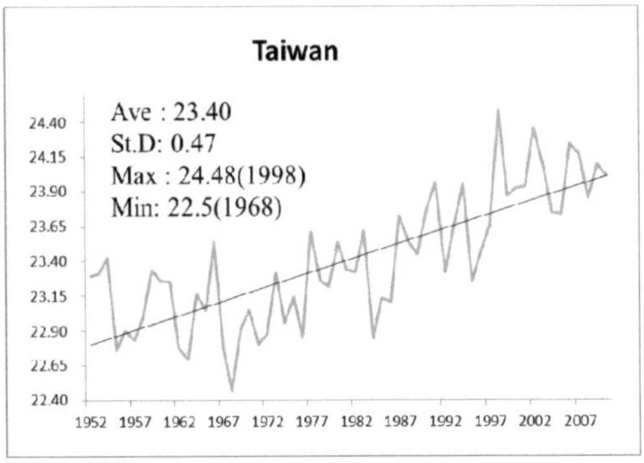

Figure 2.1. Trend of Temperature

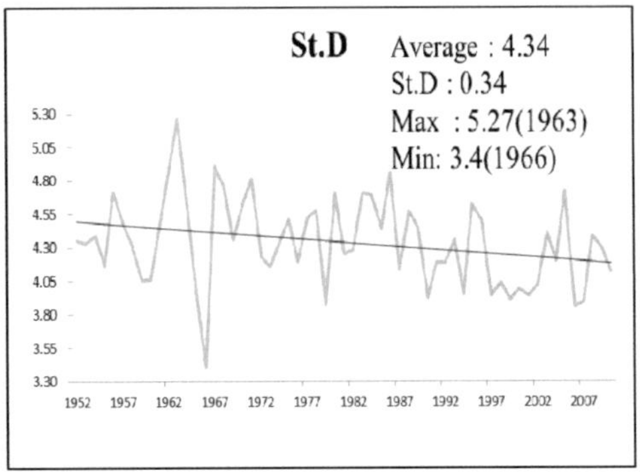

Figure 2.2. Variability of Temperature

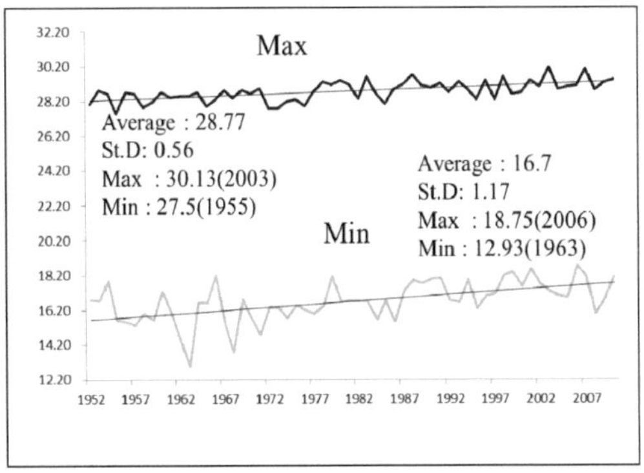

Figure 2.3. Temperature Extremes

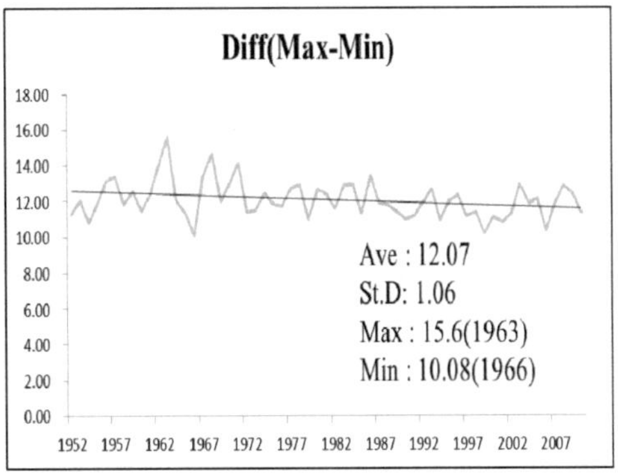

Figure 2.4. Gaps of Temperature Extremes

In addition, interesting patterns were observed among different regions and cities. The mean monthly temperature was higher in southern Taiwan than in northern Taiwan, and lower at higher altitudes than at low altitudes (Figure 3.1 – Figure 3.4). Besides, the rates of Tmax increase in rural regions (e.g., Hualien 5.09%, 1952-2010) were greater than those in urban regions (e.g., Taichung, 3.18%, 1952-2010), but the rates of Tmin increase in rural regions (e.g., Hualien 5.71%, 1952-2010) were lower than those in urban regions (e.g., Taichung, 6.88%, 1952-2010). In comparison to the urban regions, Tmax increased more significantly in rural regions. Thus, the trends of Tmax and Tmin in Taiwan were not only associated with global warming, but also with local climate change; in addition, human activities also played considerable roles. Lai and Cheng (2010) argue that urban development influences local climate by changing the land-surface characteristics. For example, compared to rural areas, an urban area is characterized by lower wind speeds, fewer hours of sunshine, lower visibility, larger turbidity and a higher daily mean temperature. It is also a fact that the amount of anthropogenic greenhouse gas concentrations has increased steadily in the past 40 years, which in turn has very likely caused the global average temperature to increase rapidly.

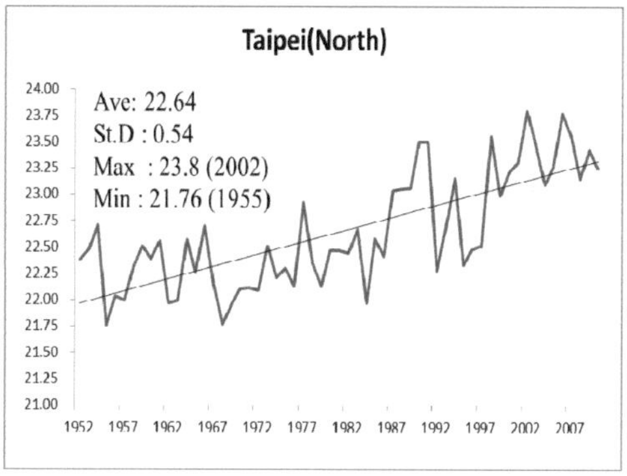

Figure 3.1. Temperature Trend in Taipei

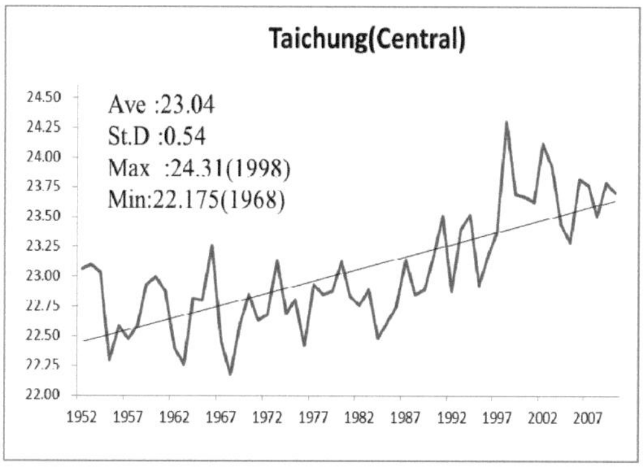

Figure 3.2. Temperature Trend in Taichung

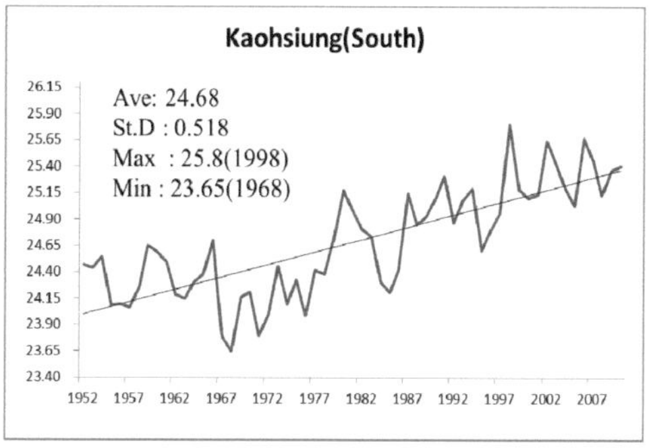

Figure 3.3. Temperature Trend in Kaohsiung

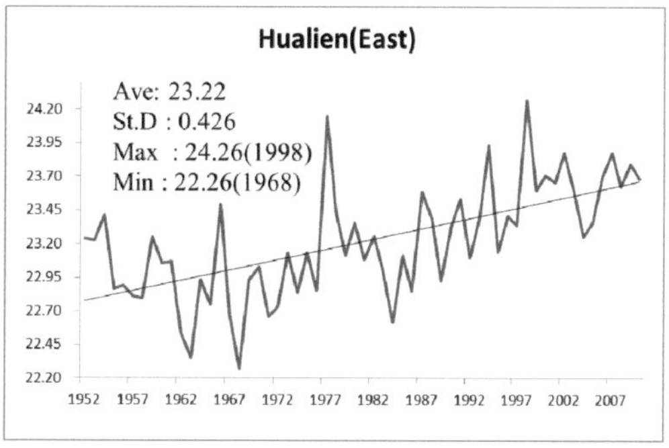

Figure 3.4. Temperature Trend in Hualien

As for precipitation, the main stream of the northward-moving Kuroshio Current passes up the eastern coast of Taiwan, thus bringing in warm and moist air. Summer and winter monsoons also bring intermittent rainfall to Taiwan's hills and central mountains. Figure 4.1 indicates that the mean monthly precipitation in Taiwan also increased from 1952 to 2010, at an average of 163.4 millimeters (mm) per month. However, unlike the trend of temperature, the variability of monthly precipitation per year has become larger (Figure 4.2). More importantly, the minimum monthly rainfall has not changed too much in the past 50 years; therefore, the variation mainly comes from the variability of the maximum monthly precipitation (Figure 4.3 – Figure 4.4).

From Figure 5.1 - Figure 5.4, we can see that the mean precipitation was much higher in northern (Taipei) and eastern Taiwan (Hualien) but lower in southwest Taiwan (Taichung and Kaohsiung). More rain falls in the mountains than in the plains, on the east coast than on the west coast, and on the windward side of hills than on the leeward (sheltered) side. The north has rain all year round while the south is rainy in summer and dry in winter. In winter, when the northeastern monsoon system is active, the north is constantly visited by drizzle while the south remains dry. However, in summer when the southwestern monsoon comes in force, afternoon thunderstorms and typhoons carry heavy rains to central and southern Taiwan. This intensive and concentrated summer rainfall, which constitutes up to 80% of annual precipitation, often causes flooding and landslides. As northern Taiwan has more rainy days than the south, the variability of rainfall slightly increases as it moves toward the south.[16]

Tropospheric ozone (O3) is a global air pollution problem and an important greenhouse gas. In large areas of the industrialized and developing world, ground-level O3 is one of the most pervasive of the global air pollutants, with significant impacts on human health, food production and the environment. Economic losses for South Asia are estimated to be in the region of US$ 3.9 billion per year for 4 staple crops (wheat, rice, soybean and potato) in Bangladesh, Bhutan, India, Nepal, Pakistan and Sri Lanka. The largest losses are found in India (US$ 3.1 billion), Pakistan (US$ 0.35 billion) and Bangladesh (US$ 0.4 billion).[17] Figure 6.1 – Figure 6.5 show the O3 concentrations in Taiwan and its four main cities from 1998 to 2010. The trend of O3 concentrations also shows increases in Taiwan. Additionally, the average parts per million (ppm) of O3 concentrations increases as it moves toward the south, whereas the least developed eastern area, Hualien, has the lowest O3 concentrations.

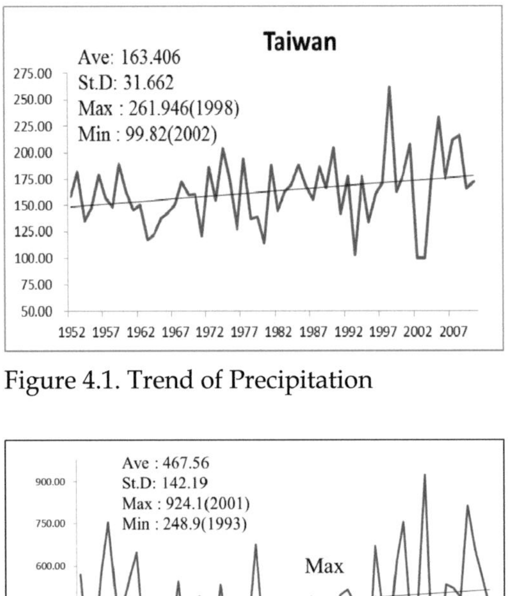

Figure 4.1. Trend of Precipitation

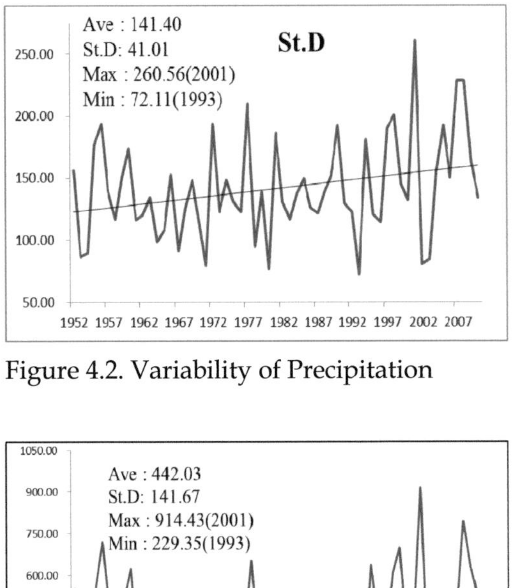

Figure 4.2. Variability of Precipitation

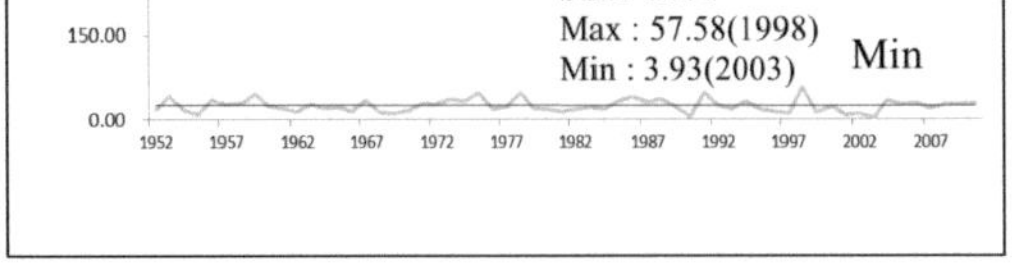

Figure 4.3. Precipitation Extremes

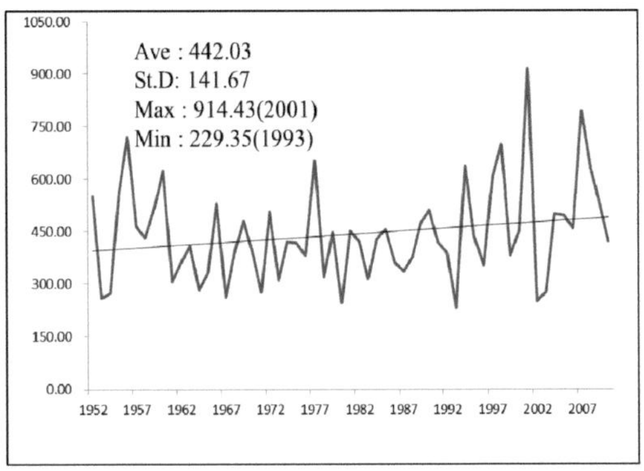

Figure 4.4. Gaps of Precipitation Extremes

[16] http://twgeog.geo.ntnu.edu.tw/english/climatology/climatology.htm

[17] http://seiinternational.org/mediamanager/documents/Publications/Climate/food_security_ozone_climate_policybrief.pdf.

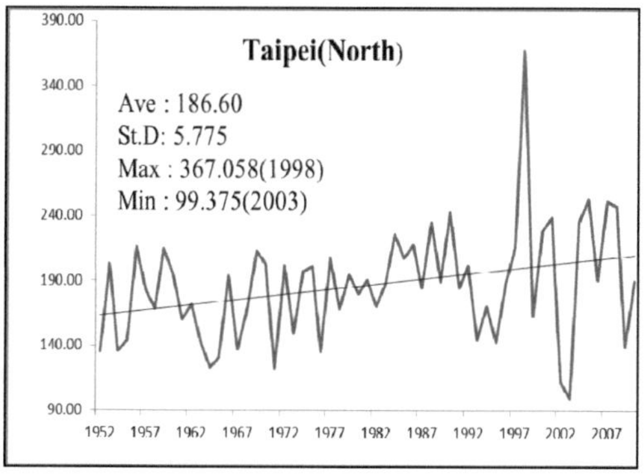

Figure 5.1. Precipitation Trend in Taipei

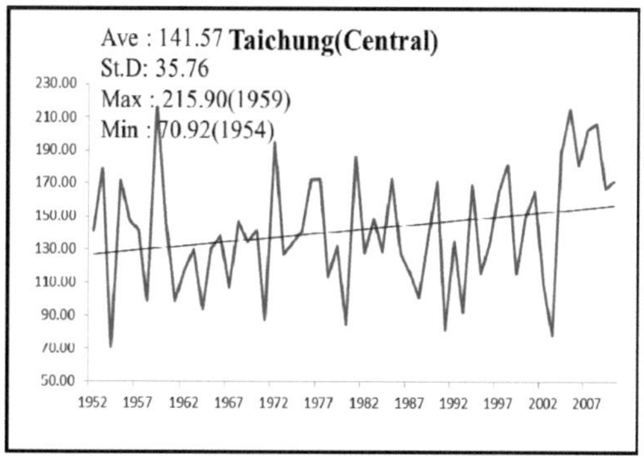

Figure 5.2. Precipitation Trend in Taichung

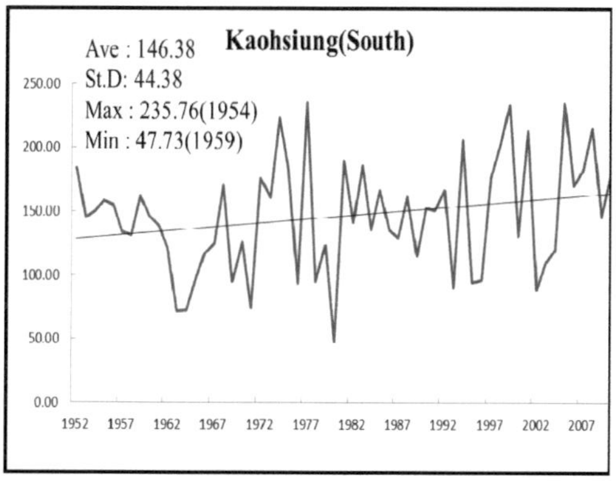

Figure 5.3. Precipitation Trend in Kaohsiung

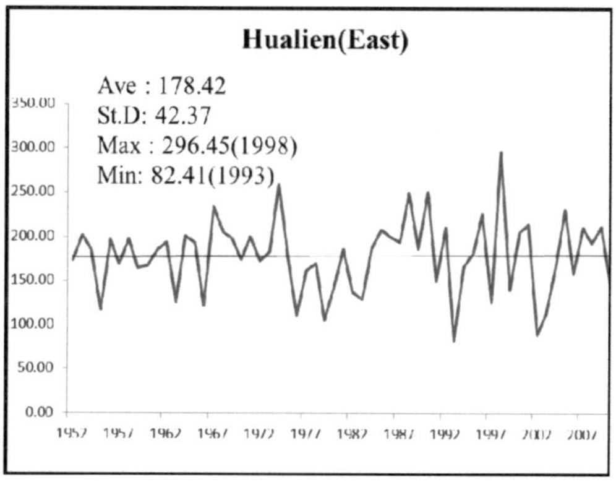

Figure 5.4. Precipitation Trend in Hualien

The correlation analyses of environmental changes and economic growth are presented in Table 3 – Table 7. From Table 3, we can find that the correlation coefficient of temperature and economic growth is negative (-0.469), implying that these two variables are negatively correlated. It still remains the same if we only focus on the agricultural sector (-0.354) or the manufacturing sector (-0.226). However, the variability of temperature shows positive values as well. As we mentioned earlier, since the variation of temperature becomes smaller as time goes by, it could be the reason why we observe this outcome. If we look at the results in different areas, with the exception of Taichung city, the three other main cities also have an inversely correlated relationship between temperature and economic growth (Table 4).

	Overall Economy (1952-2010)	Agricultural, forestry, fishery and husbandry sectors (1982-2009)	Manufacturing sector (1982-2009)
Temperature	-0.469	-0.354	-0.226
Variability (St. D)	0.212	0.453	0.193

Table 3. Correlation Coefficients of Temperature and Economic Growth in Taiwan

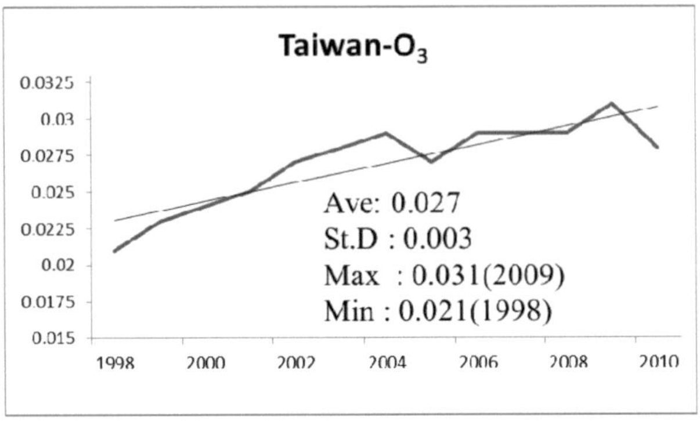

Figure 6.1. Trend of Ozone Concentrations

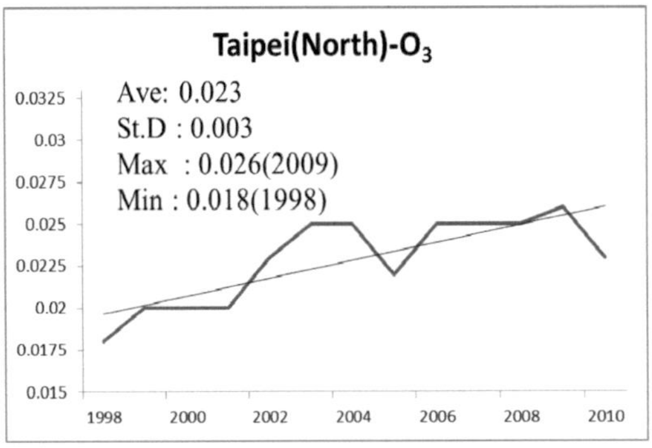

Figure 6.2. Trend of Ozone in Taipei

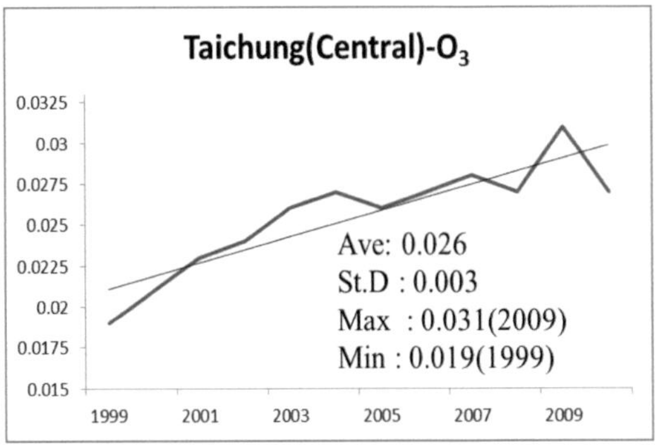

Figure 6.3. Trend of Ozone in Taichung

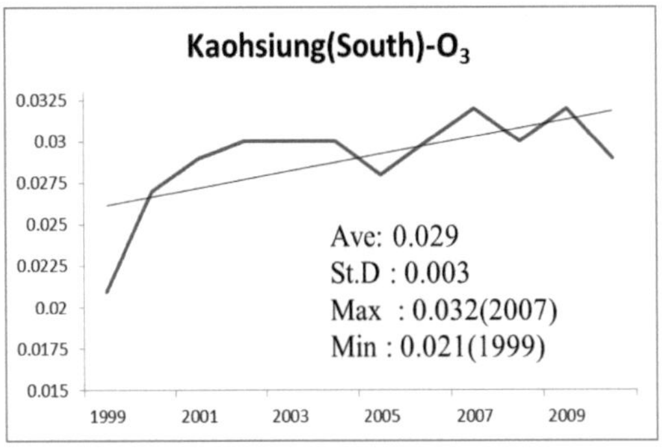

Figure 6.4. Trend of Ozone in Kaohsiung

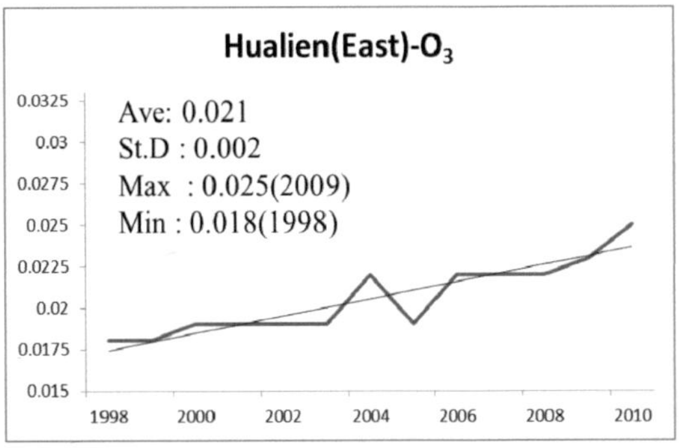

Figure 6.5. Trend of Ozone in Hualien

	Taipei (North)	Taichung (Central)	Kaohsiung (South)	Hualien (East)
Temperature	-0.061	0.118	-0.214	-0.309
Variability (St. D)	0.117	-0.132	0.053	-0.234

Table 4. Correlation Coefficients of Temperature and Economic Growth in Four Major Cities
Note: The study periods are 1999-2009.

On the other hand, the correlation coefficients of precipitation have some differences from those of temperature analyses. First, in Table 5, the correlation coefficient of precipitation and economic growth is negative (-0.365), implying precipitation level is negatively correlated to economic growth; and it still remains a negative sign in the agricultural sector (-0.343) and the manufacturing sector (-0.281). Second, variability of precipitation is inversely correlated to economic growth (-0.42), indicating that higher variability of rainfall is more likely to come with lower economic growth. Third, except for Kaohsiung City (south), the results still indicate that higher precipitation level and variability are negatively related to economic growth in three other major cities in Taiwan (Table 6). Furthermore, the correlation analyses of ozone concentrations are shown in Table 7. It also demonstrates that ozone level and economic growth have negative correlation in Taiwan, and in the four different major cities.

	Overall Economy (1952-2010)	Agricultural, forestry, fishery and husbandry sectors (1982-2009)	Manufacturing sector (1982-2009)
Precipitation	-0.365	-0.343	-0.281
Variability (St. D)	-0.42	-0.376	-0.429

Table 5. Correlation Coefficients of Precipitation and Economic Growth in Taiwan

	Taipei (North)	Taichung (Central)	Kaohsiung (South)	Hualien (East)
Precipitation	-0.157	-0.176	0.154	-0.03
Variability (St.D)	-0.147	-0.369	0.129	-0.19

Note: The study periods are 1999-2009

Table 6. Correlation Coefficients of Precipitation and Economic Growth in Four Major Cities

	Taiwan	Taipei (North)	Taichung (Central)	Kaohsiung (South)	Hualien (East)
O3	-0.195	-0.241	-0.083	-0.017	-0.018

Note: The study periods are 1999-2009.

Table 7. Correlation Coefficients of Ozone and Economic Growth in Taiwan

From the correlation analyses above, we can conclude that the temperature, precipitation and ozone concentrations are inversely correlated to economic growth in Taiwan. In addition, not only the level, but also their variability has the same inverse correlation to economic growth.

5. Conclusion

Understanding the impacts of environmental changes on the nation's economy is, obviously, of paramount importance for both the mitigation of environmental change and the development of adaptation policies. This is a preliminary study to investigate how the environment in Taiwan has changed in the past years, and how it is related to economic growth in Taiwan. This study, first, collects a set of long-term data of temperature, precipitation and ozone concentrations in Taiwan and its four major cities. By observing the data, we find that the temperature in Taiwan significantly increased during the period of 1952-2010; however, the temperature variability within a single year becomes smaller. On the other hand, the mean monthly precipitation in Taiwan also increased during the same period, but unlike the temperature, the variability of monthly precipitation becomes larger. More importantly, the minimum monthly rainfall has not changed significantly in the past 50 years; therefore, the variation mainly comes from the variability of the maximum monthly precipitation. Besides, the trend of O3 concentrations also shows an increase in the past years, and the O3 level increases as it moves toward the south in Taiwan.

The correlation analyses demonstrate that the level of temperature, precipitation and ozone concentrations are negatively correlated to economic growth in Taiwan. Furthermore, the variability of precipitation is also found to be inversely related to the economic development, but we do not find the same result for temperature variability. Therefore, from this study, although environmental changes are shown to be inversely related to economic growth in Taiwan, the temperature, precipitation and ozone concentrations seem to have their own different ways to affect economic growth, not only from their level but from their variability. In order to investigate how and to what extent these environmental changes influence the economic development in Taiwan, a more comprehensive econometric model is suggested in the future.

6. References

[1] Aggarwal, P. K. (2003)Impact of Climate Change on Indian Agriculture, Journal of Plant Biology, Vol.30,No.2, pp.189–198.
[2] Bächler, G. & K. R. Spillmann (1996). Environmental Degradation as a Cause of War, Regional and Country Studies of Research Fellows. Zürich: Verlag Rüegger.
[3] Deschenes, O. & M. Greenstone (2007). The Economic Impacts of Climate Change: Evidence from Agricultural Output and Random Fluctuations in Weather, American Economic Review, Vol.97, No.1, pp.354-385.
[4] Dell, M., B. F. Jones & B. A. Olken (2008). Climate Shocks and Economic Growth: Evidence from the Last Half-Century, , National Bureau of Economic Research, No.14132.
[5] Dell, M., B. F. Jones& B. A. Olken (2009). Temperature and Income: Reconciling New Cross-Sectional and Panel Estimates, American Economic Review Papers and Proceedings, Vol. 99, No.2, pp.198-204.

[6] Homer-Dixon, T. & P. Gizewski (1998). The Case of Pakistan, in Homer-Dixon, T., Blitt, J., (Eds), Ecoviolence: Links Among Environment, Population, and Security. Rowman and Littlefield, Lanham, pp.147-200.
[7] Homer-Dixon, T. (1999).Environment, Scarcity, Violence, Princeton, NJ: Princeton University Press.
[8] Hou, J. (2000)From Dual Disparities to Dual Squeeze: The Emerging Patterns of Regional Development in Taiwan Berkeley Planning Journal, Vol.14, No.1, pp.4-22.
[9] House of Lords (2005). The Economics of Climate Change, HL Paper 12-I, Select Committee on Economic Affairs 2nd Report of Session 2005-06, London.
[10] Hsu H. H. & C. T. Chen (2002). Observed and Projected Climate Change in Taiwan.Meteorology and Atmospheric Physics, Vol. 79, No. 1, pp. 87-104.
[11] Koubi, A, T. Bernauer, V. Kalbhenn,& G. Ruoff (2010), Climate Change, Economic Growth, and Civil Conflict, Climate Change and Security, pp.21–24
[12] Lomborg, B. (2001).The Skeptical Environmentalist: Measuring the Real State of the World, Cambridge: Cambridge University Press.
[13] Lai, W. L. & W. L.Cheng (2010). Air Temperature Change Due to Human Activities in Taiwan for the Past Century, International Journal of Climatology, Vol. 30, No. 3, pp. 432–444.
[14] Mendelsohn, R., W. Morrison, M. Schlesinger & N. Andronova (1998). Country-Specific Market Impacts of Climate Change , Climate Change, Vol. 45, No.3, pp. 553-569.
[15] Mendelsohn, R., A. Dinar & L. Williams (2006).The Distributional Impact of Climate Change on Rich and Poor Countries, Environment and Development Economics, Vol.11, No.2, pp. 159-178.
[16] Mendelsohn, R. (2009). Climate Change and Economic Growth, World Bank Working Paper 60.
[17] Nordhaus, W. & J. Boyer (2000).Warming the World: The Economics of the Greenhouse Effect, Cambridge, MA: MIT Press.
[18] Roson, R. & D.V. Mensbrugghe (2010). Climate Change and Economic Growth: Impacts and Interactions , Department of Economics, University Ca' Foscari of Venice Research Paper Series, No. 2010_07.
[19] Stern, D. I. (1996). Economic Growth and Environmental Degradation: The Environmental Kuznets Curve and Sustainable Development, World Development, Vol. 24, No.7, pp.1151-1160.
[20] Simon, J. L. (1998). Lebensraum: Paradoxically, Population Growth May Eventually End Wars, Journal of Conflict Resolution, Vol.33, No.1, pp.164-180.
[21] Shein, K. A., A. M. Waple, M. J. Menne,& J. C. Christy (2006). State of the Climate in 2005, Bulletin of the American Meteorological Society, 87: S6–S102.
[22] Schubert, R., H. Schellnhuber, N. Buchmann, A. Epiney, R. Grießhammer, M Kulessa, D. Messner, S. Rahmstorf & J. Schmidt (2009). Climate Change as a Security Risk, German Advisory Council on Global Change (WBGU) Flagship Report. Earthscan Publications Ltd.: London.
[23] Tol, R. (2002). Estimates of the Damage Costs of Climate Change-Part II: Dynamic Estimates, Environmental and Resource Economics, Vol. 21, No.2, pp.135-160.

6

Low-Carbon Pilot Tour and Municipal Government Investment: Taiwan's Experience

Liang-Feng Lin
Department of Accounting, National Chengchi University
Taiwan

1. Introduction

In the 1980s the carbon-mitigation strategy of Taiwan focused on the recycling of community waste (Chang and Chang, 1993). In the 1990s and 2000s the strategy shifted to reducing the Green House Effect and introducing low-carbon energy technologies such as wind power, solar power, gas, biomass energy, and other reusable energies. Many policies tried to encourage businesses to development reusable energies such as tax deductions or subsidies. Many researches attempted to evaluate the policy effectiveness of the reduction of the various types of pollution, and most of results found that the reduction effects were not as good as the policy claimed or expected (Lee, 1996, Lee, 2003,Wang, 1999). In recent years, the government has tried to embed green construction, green life and green transportation into people's lives.

The first low-carbon region case developed by the central competence government is Penghu Island. The Ministry of Economics in Taiwan announced a project in November 2010 that Penghu Island was to be established as a low-carbon island and would become a worldwide benchmark of a low-carbon island. The project expects to be completed in 2015. The estimated achievements include (1) reusable energy comprising 56% of all energies, which will more than Penghu's total non-reusable power, and (2) CO_2 emissions will be 50% less than the 2005s level, which averaged 2.1 tons per person a year (Council of Economic Planning and Development, 2010).

Penghu was chosen as a low-carbon island because many researchers found that Penghu to be an excellent place to develop wind power (Lee, 2003, Taipower, 1996, Lin, 2010, Council of Economic Planning and Development, 2010). The government is going to invest 8 billion NT dollars (€20 million). However, Penghu is just starting the process, and no solid results have yet emerged.

The idea of a low-carbon energy region in Taiwan is quite new and undeveloped. Most of the studies related to low-carbon life or area assessments focus on carbon emission reduction (Lee and Huang, 2004, Chang, 2008), and the effectiveness of different types of pollution control (Chen, 1992, Huang, Lee, and Jhuang, 2007). Many studies also attempt to provide policy suggestions (Lin, 2010, Lee, 1995). Studies related to Penghu as a low-carbon island focus on either technical skills (Lee 2003, Lee and Huang, 2004, Huang 2009, Huang, Lee and Zen, 2006) or related policy solutions (Wang, 2006, Huang).

On the other hand, New Taipei City[18] is the first municipal government to promote a low-carbon pilot tour in its jurisdiction districts. Pinglin District, a declining mountain city under the jurisdiction of New Taipei City, was the first city to promote a low-carbon pilot tour (LCPT) in 2008. The next year Shuangxi District, a long ignored coal-mining city adjacent to Pinglin, also introduced a similar pilot tour. After promoting such tours for several years, both cities started to enjoy the benefits of the LCPT. The benefits include an increasing number of tourists, economic growth, increased job opportunities and a reduction of carbon emissions. Unlike Penghu's low-carbon island project with billions (NT$) in investments, the New Taipei City Municipal Government invested very little to achieve a great performance. Therefore, the paper tries to analyze how the low-carbon tours work for both cities in relation to the infrastructure and investments of the municipal city to design and promote the tours. The success of LCPT not only has revived declining cities but has also been effective in raising the awareness of local residents and nonlocal tourists of environmental sustainability while inculcating the idea that creating a green city does not require a huge investment on the part of the municipal city government. The rest of the chapter is organized as follows. Sections II and III introduce Pinglin District and Shuangxi District, low-carbon pilot tours and the benefits from the tours. Section IV concludes the paper.

2. Pinglin District and the low-carbon pilot tour

2.1. Area introduction

Pinglin District, a 173.83 square kilometer town, is the third largest district of New Taipei City. The town used to be a very important midpoint between Taipei and Yilan going back to the Ming Dynasty. Although mountain ridges in this area resulted in twisting and winding roads, Pinglin still served as an important hub to deliver products from the eastern part to the western part of Taiwan. In addition to its important transportation position, Pinglin is also famous for its Wensan Tea and its beautiful views. However, after 2006, the opening of Highway 5 connected Yilan and Taipei to full traffic, a reduction of travel time from Yilan to Taipei of almost 3 hours. The highway suddenly changed the city from a major hub city to a minor tourist attraction. Pinglin is located upstream of Peishi Creek, the residential water source for the Taipei area; therefore, its city constructions have been subject to restrictions since 1987. The declining economy, a decreasing number of tourists, and limited construction opportunities forced many young people go elsewhere to find better jobs.

2.2. The Low- carbon pilot tour in Pinglin

On 2006 June 16, just the second day after Highway 5 opened to full traffic, the once busy traffic of Pinglin suddenly became quiet. As the old saying goes, "a crisis is a turning point." Pinglin is a prime example. In 2008 December 19th, New Taipei City and the Pinglin District Municipal Government (PDMG) designed the first new carbon pilot tour in Taiwan and officially established Pinglin's Low-Carbon Tour Service Center (Wu, 2008). The service

---

[18] New Taipei City was formerly referred to as Taipei County which was renamed as New Taipei City since 2011. In order to simplify the description, the paper uses New Taipei City and does not use Taipei County. The same rule applies to Pinglin and Shuangxi.

center tries to combine the promotion of a low-carbon lifestyle with a city tour. It helps local stores to support low-carbon operations. The famous tea industry and its distinctive scenery of mountains and creeks give Pinglin tremendous potential to promote the tourism industry. The low-carbon tour allows tourists to enjoy the beautiful natural local scenery and the town's ample cultural assets; at the same time, it also helps them understand the ecological environment of Pinglin. The most important thing is that the tour boosts the local economy without damaging the local environment.

According to estimates of the Environmental Protection Agency of New Taipei City, a trip from the city to Pinglin creates at least 23 kilograms of carbon per person[19]. Based on the above idea, the low-carbon pilot tour in Pinglin was designed as follows:

1). Walk, bike and commute: The PDMG designed the tour for every Wednesday and Saturday. On Highway 9 at 35.5 kilometer a road block has been set up; only buses with 10 or more persons are allowed to enter the district. Upon entering the district, the PDMG provides a free transfer bus to the Pinglin Tea Industrial Museum and to the entrance of the Jingualiao Community, the major scenic site of the tour. The Jingualiao Community extends along the Jingualian Creek is provided with a fish and ferns observation path where visitors are only allowed to walk, bike, or take the 14-seat electric motorcar. Local tour guides and ecological narrators relate the history and humanities of Pinglin.

2). Seasonal and local foods: The PDMG encourages tourists to consume items from green mark stores (low-carbon stores). The stores provide only seasonal and local foods; styrofoam disposable dishware and plastic bags are prohibited.

3) Carbon coupon: The PDMG also came up with the idea of a carbon coupon in 2009. Consumers use the coupon at green mark stores, whereupon the stores donate 10% of their sales to the "Tree Planting Fund" (TPF). The concept of the carbon coupon comes from the zero carbon emission tour. According to estimates, by taking the public commuting transportation to Pinglin , walking and biking around the district, then purchasing items at green mark stores, a tourist can approximately reduce about 19 kilograms of carbon emissions per person. In order to neutralize the full 23 kilograms of carbon emissions on the tour, the extra 4 kilograms of reductions will come from planting trees. Therefore, the Pinglin low-carbon pilot tour has also been labeled the "zero carbon tour," since the TPF will use donations to plant trees in certain areas in Pinglin to reduce carbon emissions.

4). There are other low-carbon designs for pilot tours. To further reduce carbon emissions, in 2011 the PDMG built the first solar energy environmental public lavatory in Taiwan. The lavatory uses 6 solar photovoltaic cells, with maximum power generation of 1200 KWh. The estimated carbon reduction is 756 kilograms per year (Chen, 2011). Starting from the current year of 2011, the tour decided to provide a "free electric motor bike service" to offer tourists a free test ride on an electric motor bike to tour the city. Besides the free test ride, the PDMG also built an electric motor bike recharging station, which uses solar photovoltaic technology to generate power. The station is free to encourage those who use electric motor bikes to visit Pinglin.

---

[19] Huang, J. T., 2011, Card News, http://www.cardu.com.tw/news/detail.htm?nt_pk=27&ns_pk=10724

2.3 Investment by the municipal government

The total Final Accounts for Pinglin District was NT$ 564.13 million (13.43 million) in 2010 (New Taipei City Final Accounts, 2010), which ranks 20th of the 29 districts in New Taipei City. However, to promote the low-carbon pilot tour, New Taipei City invested very little in related activities. Pinglin has been the water source for Taipei since 1987, and since then the district has seen limited construction projects. All riverside constructions and bike trials use ecological engineering methods, most of which were completed before 2008. The budget for ecological engineering was NT $15 million (0.36 million) per year (Wu, 2009). The largest single investment was the solar energy public lavatory which cost NT$ 20 million. Because of the success of the LCPT, Pinglin won the gold medal for environmental demo community in 2010, and received subsidies from the Environmental Protection Administration. The PDMG received about NT$7 to NT$9 million in subsidies for promoting the low-carbon pilot tour from the New Taipei City Municipal Government. Table 1 lists the related investments. The table indicates that the total expenditures of the LCPT comprised approximately 4.5% of total budget expenditures of Pinglin, which in turn only comprised about 1.21% of New Taipei City's total expenditures for 29 districts.

Unit: NT Thousand

Items	Final Accounts	%	Note
Ecological Engineering Construction	15,000	2.66%	Per Year
Low-Carbon Pilot Tour	8,000	1.42%	Per Year
Solar Energy Public Toilet	2,000	0.35%	2011
Other Operating Expenses for LCPT	100	0.02%	Per Year
Pinglin Final Accounts for 2010	564,135	100.00%	
New Taipei City Final Accounts for 2010*	76,507,090		

*Pinglin District ranks 20th of 29 districts of New Taipei City in total expenditures and its budget comprised only 1.21% of total expenditures.

Table 1. Pinglin District Pilot Tour Related Expenditures in 2010

2.4. Benefits from the LCPT

The Environmental Protection Administration of New Taipei City found that the LCPT created the following benefits for Pinglin. The tour attracted 25,460 persons in 2008, and approximately 400,000 persons in 2009 and 2010 in total (with an average of 25,091 per month). In 2008 the tour created NT$7 million in economic benefits and NT$410 million in 2009 and 2010 in total. In 2008 Pinglin District hired 49 staff members for the activities and 80% of them are from the local community; in 2009 and 2010 the tour provided 52 job opportunities for local citizens. In 2008 carbon emission reduction was 48,762 kilograms (equivalent to the planting of 10,000 trees a year), but the number increased to 7,536,248 kilograms (equivalent to planting1,545,027 trees a year) in 2009 and 2010. Table 2 summarizes the benefits from the LCPT (New Taipei City Low Carbon Life Net, 2011).

Pinglin District also was awarded by the Environmental Protection Demo Community from Environmental Protection Administration and NT$2,000 in subsidies. Besides those tangible benefits, the biggest intangible gains are the increased awareness of environmentalism of the residents, according to District Major Chao-Qin Wang. Within three years the residents have changed from resisting the tour to supporting the tour; from environmental skeptics to green-life believers.

Items[a]	2008	2009 and 2010 in Total
Tourists	25,460 persons	400,000 (25091 per month)
Economic Benefits	NT $7,000,000	NT $410,000,000
Job Opportunities	39	52 each year
Carbon Emission Reductions	48,726 Kg	7,536,248 Kg
	10,000 TPE[b] a year	1,545,027 TPE a year

a The first year numbers only include those who registered for LCPT, and the other years also include self-guided tourists.
b Tree Planting Equivalent

Table 2. Summary of Benefits from Pinglin LCPT

3. Shuangxi District and its low-carbon pilot tour

3.1. Area introduction

Shuangxi District means "two creeks"-- Pinglin creek and Mudan creek, both of which pass through the city. The district is located on the outskirts of Taipei City. The natural environment is not affected by over-development of the metropolitan area, and therefore the district still has ample green resources. The district used to be famous for coal mining in the 1970s, but the industry had been in decline for a long time, and the population decreased from 50, 000 in 1970 to 9,000 in 2006. Like Pinglin, Shuangxi has also been affected by the construction of Highway 5. After the highway opened to traffic in 2006, Shuangxi became even less important than before. However, the low development of the district also became a turning point for developing Shuangxi as an eco-city and a green tourism district. Shuangxi completed a loop bike trail and constructed a bicycle rental station to serve as a healthy walking path and bike riding along the ancient trial of Ma-Zhu Keng in 2008. Shuangxi began introducing "Low-Carbon Pilot Tours" from Pinglin in 2009.

3.2. Low-carbon pilot tour design

Unlike Pinglin, which, as a relatively isolated city could only be reached by Highway 9, Shuangxi is accessed by highways 2C, 102, 106 and a railroad. The tour is designed as follows:

1). Two Creeks, Two Irons and Low-Carbon Tours: The tour slogan is "Two Creeks, Two Irons and Low-Carbon Tours". Two Creeks refers to the name of Shuangxi. Two Irons refers to a railroad trip to Shuangxi and a bike ride here to reduce carbon emissions. Those wanting to go on such a tour need to register on the Internet. Tour capacity is 202 persons. The Shuangxi District Municipal Government (SDMG) integrates the railway and the bike trail and holds a series of events to promote a happy and healthy low-carbon lifestyle. The local tour guides and ecological narrators combine the history and the humanities of Shuangxi District in providing visitors with an enjoyable "Healthy and Low-Carbon Tour."

2). Happy Farm. The SDMG designated some farms as Happy Farms. Within the farm, each 10 square meters is considered one unit; tourists can be adopted by a person or group to plant vegetables using an organic approach. The tourists can pick up local vegetables on a Happy Farm or purchase vegetables and fruits from local produce centers to promote the local economy.

3). Green Mark Restaurants and Stores. Tourists can have their local vegetables prepared form them at green mark restaurants. In these restaurants, tourists need to prepare their own dishware. (styrofoam disposable dishware and plastic bags are prohibited). All food waste from restaurants are recycled as compost. Similar to Pinglin, the green mark stores only sell local products, do not provide plastic bags, and strictly require garbage to be recycled.

4). The Shuangxi Low Carbon Tour Center provides tourists with low-carbon tour program planning, answer inquires on the bike and electric motor rental service station, and consulting services for low-carbon meals.

3.3. Municipal government investment

The total final account for Shuangxi District was NT$405.164 million (€9.647 million) in 2010 (New Taipei City Final Accounts, 2010), which ranked 25th of the 29 districts of New Taipei City. The Shuangxi District started building an eco infrastructure in the years from 2006 to 2008. The loop bike trial and healthy walking path along with the ancient trial of Ma-Zhu Keng were completed during that period. The New Taipei City District Government provided only NT$ 15 million per year for each district. The "Two Creeks, Two Irons and Law-Carbon Tours" received around NT $700 million in subsidies from the City Municipal Government per year. Therefore, the tour-related expenditures were minimal. Table 3 indicates the related expenditures. The tour-related expenditures comprised only 5.45% of total expenditures, which comprised only 0.53% of New Taipei City's total expenditures for 29 districts.

Unit NT Thousand

Items	Final Accounts	%	Note
Ecological Engineering Construction	15,000	3.70%	Per Year
Low-Carbon Pilot Tour	7,000	1.73%	Per Year
Other Operating Expenses for LCPT	80	0.02%	Per Year
Shuangxi Final Accounts for 2010*	405,164	100.00%	
29 Districts Total Final Accounts for 2010	76,507,090		

*Shuangxi ranks 25th of 29 districts in total expenditures and only comprised 0.53% of the total expenditures of Taipei City.

Table 3. Shuangxi District Pilot Tour Related Expenditures in 2010

3.4. Benefits from the LCPT

The Environmental Protection Administration of New Taipei City found that the LCPT created the following benefits for Shuangxi. The tour attracted 6,280 tourists in 2009. and 7,509 persons in 2010(Environmental Protection Administration, 2010). According to the Tourism Bureau's estimation, each tourist spent on average NT $1,019 per day in Taiwan; therefore, the LCPT of Shuangxi promoted NT $6,949 thousand and NT $7,652 thousand in economic benefits for the years 2009 and 2010, respectively. The tour created 15 job opportunities in 2009 and 19 in 2010. Carbon emissions were reduced by 44,807 Kg and 54,673 Kg in 2009 and 2010, respectively, which is equivalent to planting 9,188 and 12,095 trees a year, respectively. Table 4 summarizes the benefits derived from the tours.

Items	2009	2010
Tourists[c]	6,820 persons	7,509 persons
Economic Benefits (thousand)	NT $6,949	NT $7,652
Job Opportunities	15	19
Carbon Emission Reductions	44,807 Kg	54,673Kg
	9,188 TPE a year	12,095 TPE a year

c. The number of tourists includes only those who registered for the LCPT; unregistered tourists are not included. The same is true for other statistics presented in the table.

Table 4. Summary of Benefits from Shuangxi LCPT

4. Conclusion

In 2008 Pinglin, a mountain city in decline, became the first city in Taiwan to promote a low-carbon pilot tour. The following year Shuangxi, a long ignored coal mining city located near Pinglin, also introduced similar activities. After promoting such tours, both cities immediately started to enjoy the benefits of a low-carbon life style. The benefits include a growing tourism, improved economic growth, an increased number of job opportunities and a reduction of carbon emissions. Unlike Penghu's low-carbon island project with its billions dollars of investments, the New Taipei City Municipal Government invested very little to achieve a solid performance. Pinglin and Shuangxi, in promoting low-carbon pilot tours, not only have effectively turned around their declining economies but are also contributing to the sustainability of their homeland. The tours not only teach tourists that low-carbon activities can be fun; they also help residents understand how low-carbon activities can be integrated into their lives. Moreover, from the tours, we have learned that environmental sustainability does not require major investments. Both cities made use of minimal investments to achieve maximum benefits for their local economies and environments.

Once Pinglin began promoting low-carbon pilot tours in 2008, many other cities came to offer similar tours. In 2011 the Tourism Bureau planned 20 low-carbon routes in coordination with the Taiwan Tourist Bus Travel Service. The travel service is designed for tourism planning and the design of bus services. From the major attraction of the Taiwan Train Station, a high-speed train shuttles passengers to popular tourist attractions in Taiwan. This is a positive response to a new low-carbon life.

5. References

[1] Chang, Y. C., 2008, Low-Carbon Community: New Sustainability Development Indicators, Tai Ta E-News, January, http://www.delta-foundation.org.tw/epaper/080115/epaper_080115.htm#a1

[2] Chang, Z., and N. Chang, 1993, Evaluation of the Local Enviornmental Protection Administritation Waste Recycleing Performance, Environmnental Protection Administration, 154 Research Paper.

[3] Chen, W. S., 1992, Blue Sky Again, Pullution Control Status, and Prevention Strategy in Taiwan, Journal of Science, June, Vol 270, http://210.60.226.25/science/content/1992/00060270/0004.htm

[4] Chen, Y., 2011, "The First Public Restroom in Pinglin." http://travel.mtaipei.tw/index_m.php?ptype=news_view_from2&id=20844.

[5] Council for Economic Planning and Development, 2010, http://www.cepd.gov.tw/m1.aspx?sNo=0014489.

[6] Environmental Protetion Administritation, 2010, Two Creeks, Two Irons and Low-Carbon Tours." New Taipei City.

[7] Huang, C H., 2009, "Measurement of Energy Efficiency in Taiwan and Its Relevance to CO2 Decoupling." Invited paper presented at the Workshop on Innovations in Energy Efficiency. Stanford University, California, February 17-18.

[8] Huang, C. H. , C. Yang, H. Lin, and H. Su ,2006, "The Impact of the Oil Price Shock on CO2 Emissions and the Adoption of Renewable Energy." Paper presented at the International Conference on Regional Carbon Budgets, August 16-18, Beijing.

[9] Huang, C. H., 2007, "Double Dividend Hypothesis: Results from the Energy Tax Initiative in Taiwan." Invited speech presented at the National Institute for Environmental Studies, March 1, Tsukuba City, Japan.

[10] Huang, C. H., C. M. Lee, and F. C. Jhuang, 2007, Pollution Prevention Investment and Innovation Behavior under Carbon Emission Trading Scheme, Taiwan Economic Review, Vol. 35:1, 87-114.

[11] Huang, J., 2101/10/05, Card News, http://www.cardu.com.tw/news/detail.htm?nt_pk=27&ns_pk=10724
[12] Lee, C. J., and Huang, Y. C., 2004, Renwerable Energy Policy and Green House Effect of Industrial Developmen, Taiwan Watch Vol.6, No.2 Summer, 52-60
[13] Lee, C. M., C. H. Huang, and U. Zen , 2006. "Optimal Integration of Carbon Tax and Subsidy for Renewable Energy Development." Paper presented at the 2nd International Conference on Environmental History, November 8-10, 2006, Academia Sinica, Nankang, Taiwan.
[14] Lee, C., 2009, Energy Saving and Carbon Risk Reduction Assessment in Energy Intensive Industries in Taiwan, IDMC, 3DSA, Asia Display, Taipei International Center, Taiwan, April 27-30, 2009.
[15] Lee, C., and H. Liu , 2009, Assessment of GHGs Decoupling and Electricity Cost by Low-Carbon Energy Technology Development in Taiwan, The 15th International Joint Seminar on the Regional Deposition Processes in the Atmosphere (RDPA) and Climate Change in 2009, Taiwan/ Taipei.
[16] Lee, S., 2003, The Perspective of Renewable Energy In Penghu, Build a New Ocean Home Penghu Seminar,37~55.
[17] Lee, T., C. M. Lee, and C. Tseng, 2005, "International Emissions Trading Approach to Meet the Kyoto Protocol: The Implications of Trading Hot Air," selected paper presented at 28th Annual IAEE International Conference, June 3-6, 2005, The Grand Hotel, Taipei, Taiwan.
[18] Lee, T., C. M. Lee, and C. Tseng, 2009, International Emissions Trading Approach to Meet the Kyoto Protocol: The Implications of Hot Air, Energy Studies Review,Volume 15, Issue 2.
[19] Lee, Y. C., 1995, Environment Attitude and Environment Behavior: Theory and Empirical Test, Hu Ltd. Taipei.
[20] Lin, C. W., 2010, Green Energy Development: Establishing a Clean and Low-Carbon Island, Council of Economic Planning and Development News, http://www.cepd.gov.tw/m1.aspx?sNo=0014489
[21] New Taipie City 2010 Budget, 2011, http://www.finance.ntpc.gov.tw/web66/_file/1051/upload/selfpage/30255/1-5.xls.
[22] New Taipie City Low-Carbon Life Net, 2011, http://www.epd.ntpc.gov.tw/web66/_file/1150/upload/news/lowco2/result05.html.
[23] TaiPower, 1996, Evaluation of the Penghu Wind Power Generation Plan.
[24] Wang, T., 1999, Industrial Development, Energy Utilization and National Land Planning, Cross Century National Land Planning Seminar, National Pespective Foundation, Taipei.
[25] Wu, C., 2009, The Experience of Promoting Taipei County's Ecological Engineering, http://eem.pcc.gov.tw/eemadm/files/product_1/ws_29/01.doc